Diercke Geography

For Bilingual Classes

Toolkit

Moderation:
Prof. Dr. Reinhard Hoffmann

Autor:
Dirk Reischauer

unter Mitwirkung der Verlagsredaktion

westermann

Cover:
London (UK); Sydney (Australia); New York (USA)

Mit Beiträgen von:
Michael Fink, Henriette Heß, Reinhard Hoffmann, Verena Hundertmark, Dorothee Klein, Marjana Mehltretter, Ingo Warken, Horst Weible

CARL DIERCKE (geb. in Kyritz, Landkreis Ostprignitz/Preußen) lebte von 1842 bis 1913 und war Pädagoge und Kartograph. Von ihm stammt der bekannte *Diercke-Schulatlas*, der erstmals 1883 unter dem Namen *„Schul-Atlas über alle Teile der Erde“* erschien.

Weitere Informationen finden sich im Internet: *www.diercke.de – (Info/Chronik)*

westermann GRUPPE

Druck A[1] / Jahr 2018
Alle Drucke der Serie A sind im Unterricht parallel verwendbar.
Redaktion: Lars Büttner
Bildredaktion: Susanne Guse
Umschlaggestaltung: Thomas Schröder
Druck und Bindung: westermann druck GmbH, Braunschweig

ISBN 978-3-14-**114042**-2

Contents

How to Work with the Textbooks and the Atlas

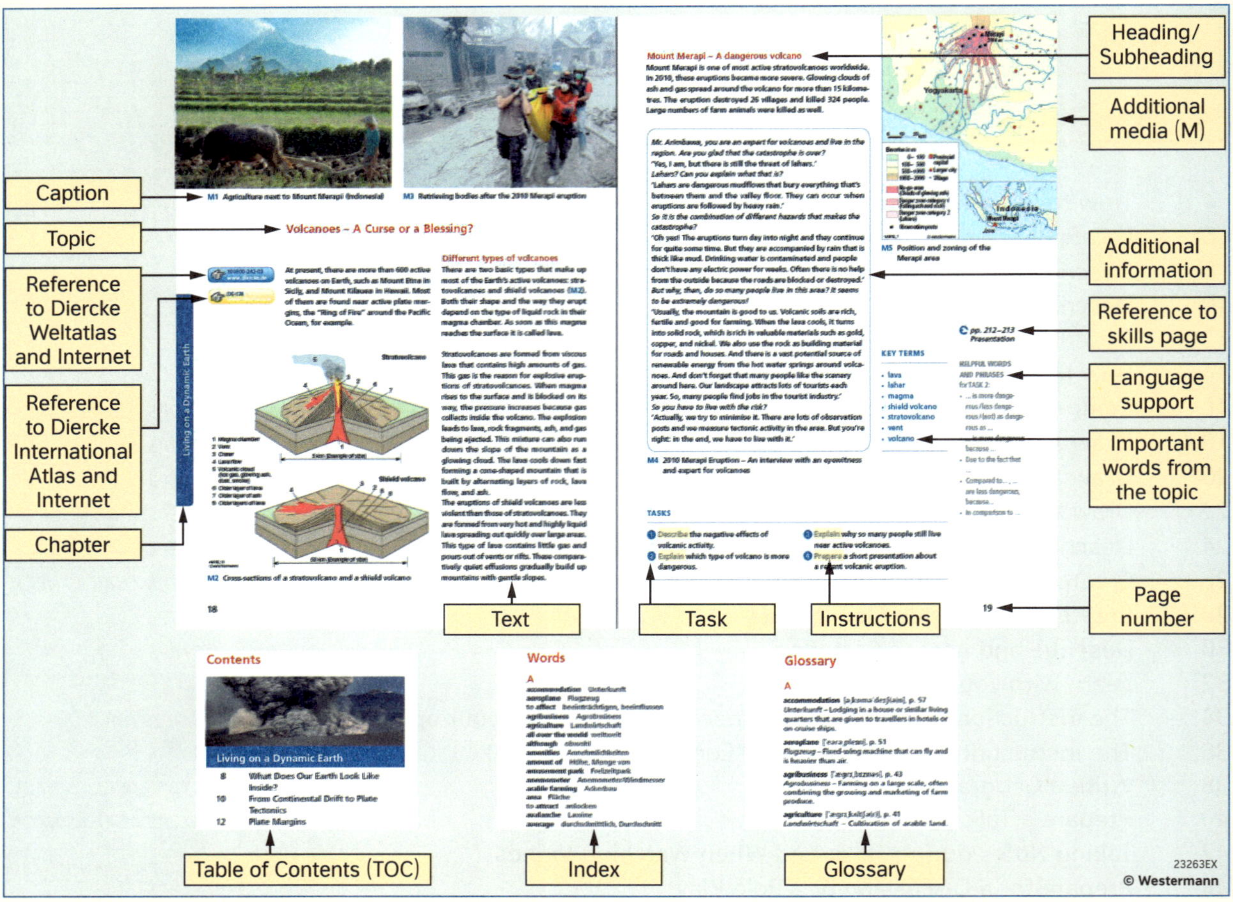

M1 Elements of the textbook (example from „Volume 1")

The two media you are likely to use most for your Geography lessons are the textbook and the atlas. They both have a table of contents, which is a list of the different chapters and topics.

At the end of textbooks and atlas you will find an index, which is a list in alphabetical order. The table of contents and the indexes are helpful tools to get information from the textbooks and atlas quickly. In the bilingual textbooks there are also German translations. The glossaries in the textbooks are lists of most KEY TERMS with translation, phonetic transcriptions and a short explanation (› pp. 52 ff).

The website *www.diercke.com* gives information for the 'Diercke International Atlas' and the 'Diercke Geography' book series in English. The information on the website *www.diercke.de* is in German.

For some of the information you need to register with the registration key (deutsch: Online-Schlüssel) from your atlas or textbook.

The textbooks

Your textbooks are an important source of knowledge (**M1**). Turning information from any textbook into knowledge is a vital skill. The SQ3R method is a very good way to help you learn efficiently (› pp. 32-33).

When you deal with the media ('**M1** ...') on a page, the sign leads you to the "Skills" pages or this Toolkit (› pp. 4-47). These pages contain useful tools to deal with all

www.diercke.de
www.diercke.com

100800-317
schueler.diercke.de

DE-204
www.diercke.com

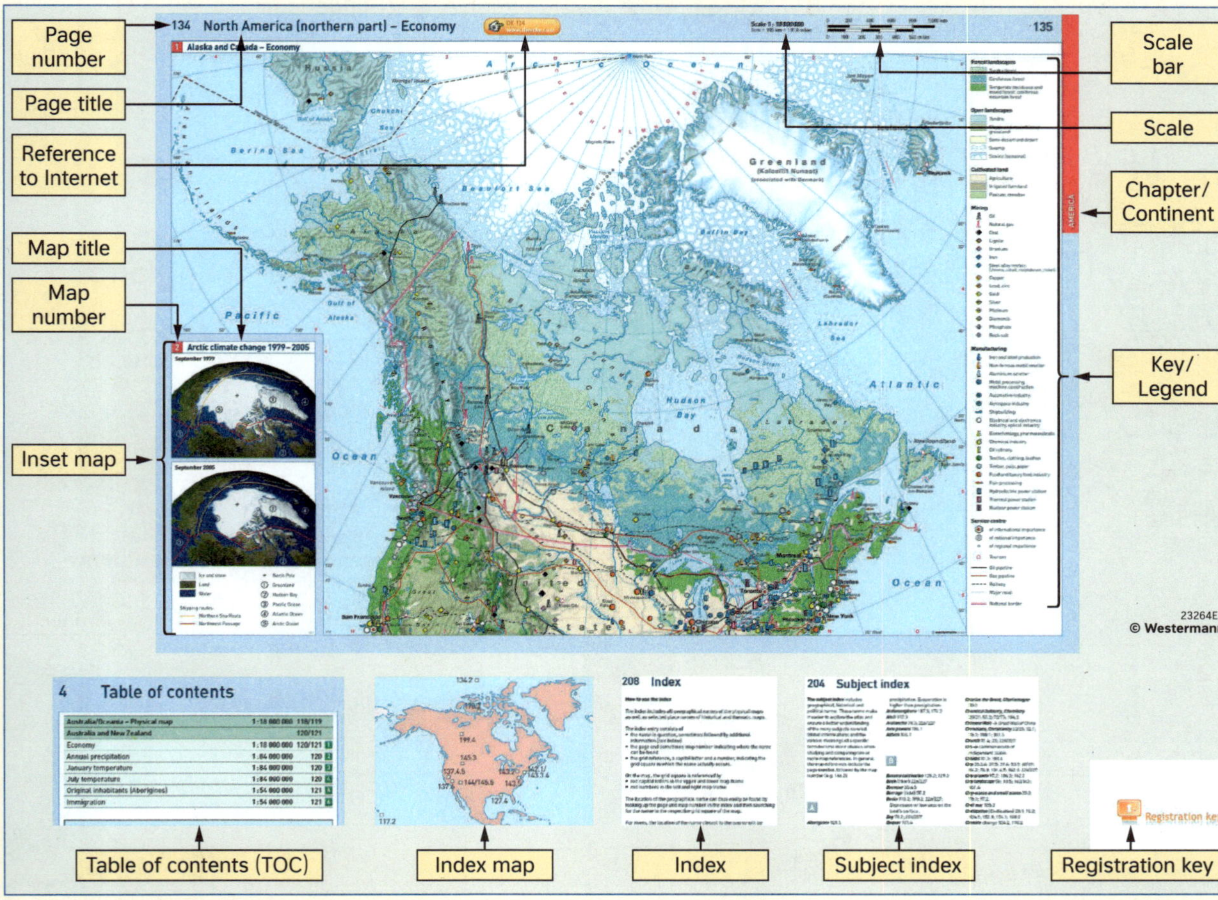

M2 Elements of the atlas

the media, often in step-by-step guides. Moreover you can find many phrases to describe these materials or to use in connection with the instructions (see below).

The TASKS are designed to access the information from the topic efficiently.
If you are not sure how to express yourself, then the 'Helpful words and phrases' on the "Skills" pages offer help.

Furthermore you will find a list of the instructions (deutsch: Operatoren) from the books with a short German explanation about what to do (› pp. 50-51). Nearly all the pages in your textbooks have a reference to an atlas and the internet.

The atlas

The atlas gives you a view of the world with its physical and thematic maps. It is full of symbols, shapes and signs. By recognizing them, geographers see the world and try to understand it.
Index maps (**M2**) will quickly inform you which maps there are about an area you are interested in. You can learn how to deal with physical maps (› pp. 8-9) and with thematic maps (› pp. 10-11). The maps in the atlas often provide additional information in inset maps or diagrams.
If you look for a certain topic in your atlas, the subject index (deutsch: Sachwortregister) is useful.
Further information about the map content is on the Internet.

Locate Places

M1 Physical map of southern British Isles

Find a place in your atlas

In order to find a place in your atlas, you can use the table of contents and look for maps that contain the place you are looking for. An easier way to do it is to use the index which lists all the places that your atlas shows in alphabetical order.

After the name of the place there is a combination of numbers and letters:

Southampton 64, C2

This entry means that you can find the city of Southampton on a map on page 64 in grid square C2 (**M1**).

100800-286
schueler.diercke.de

DE-208
www.diercke.com

The location of places on Earth

We can locate every single place on Earth with the help of only two lines (**M2**):

Lines of latitude or, simply, latitudes, run round the Earth parallel to the Equator. The Equator is at 0 degrees. It divides the Earth into the Northern Hemisphere and the Southern Hemisphere. It is the longest line of latitude (40,075 km).

The North Pole is at 90°N (90 degrees north), the South Pole is at 90°S (90 degrees south). They are the shortest lines of latitude – in fact, they are only dots.

We number the lines of latitude from 0° to 90° and we label them with the letter N for the lines in the Northern Hemisphere and with the letter S for the lines in the Southern Hemisphere.

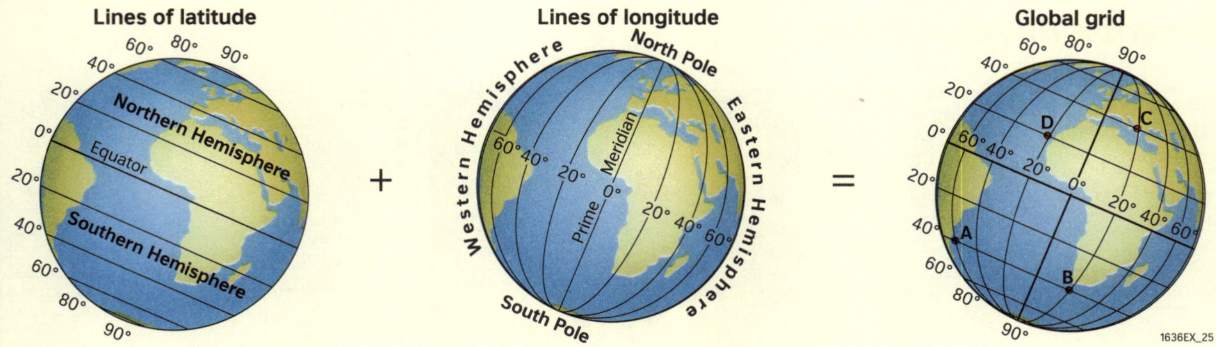

M2 The global grid

M3 The exact location of Southampton

Lines of longitude or, simply, longitudes, connect the North Pole and the South Pole. They run from north to south. Half of the circle from one pole to the other is a meridian. The Prime Meridian is at 0° and runs through Greenwich (London).
The Prime Meridian divides the Earth into the Western Hemisphere and the Eastern Hemisphere.
We number meridians from 0° to 180°. We label them with the letter W for the lines in the Western Hemisphere and with E for the lines in the Eastern Hemisphere.

Together, these two sets of lines form a grid. We call it the global grid (**M2**). The exact location of a place is where the longitudes and latitudes meet.
Most maps show the global grid as a set of thin lines with numbers, often near the edge of the map (**M3**).

You give the location of a place like this – first latitude then longitude:

Southampton 50° 54' N / 1° 24' W

You say: *Southampton is located at 50 degrees 54 minutes North and 1 degree 24 minutes West.*
The square grid of maps applies only to one specific map. The global position with longitude and latitude is universally valid.

SKILLS HACK

Degrees, Minutes and Seconds

1° (1 degree) = 60' (60 minutes)
1' (1 minute) = 60'' (60 seconds)
Be careful:
This means that 1.5° equals 1° 30'.
In the example above, 50° 54' N is the same as 50.9° N.

HELPFUL WORDS AND PHRASES
to locate a place:
"at" when the location is on the global grid
"in" when the location is in a grid square

- ... lies at/in ...
- ... is at/in ...
- ... is located at/in ...
- ... is situated at/ in ...
- The exact position of ... is at/ in ...

Describe Physical Maps

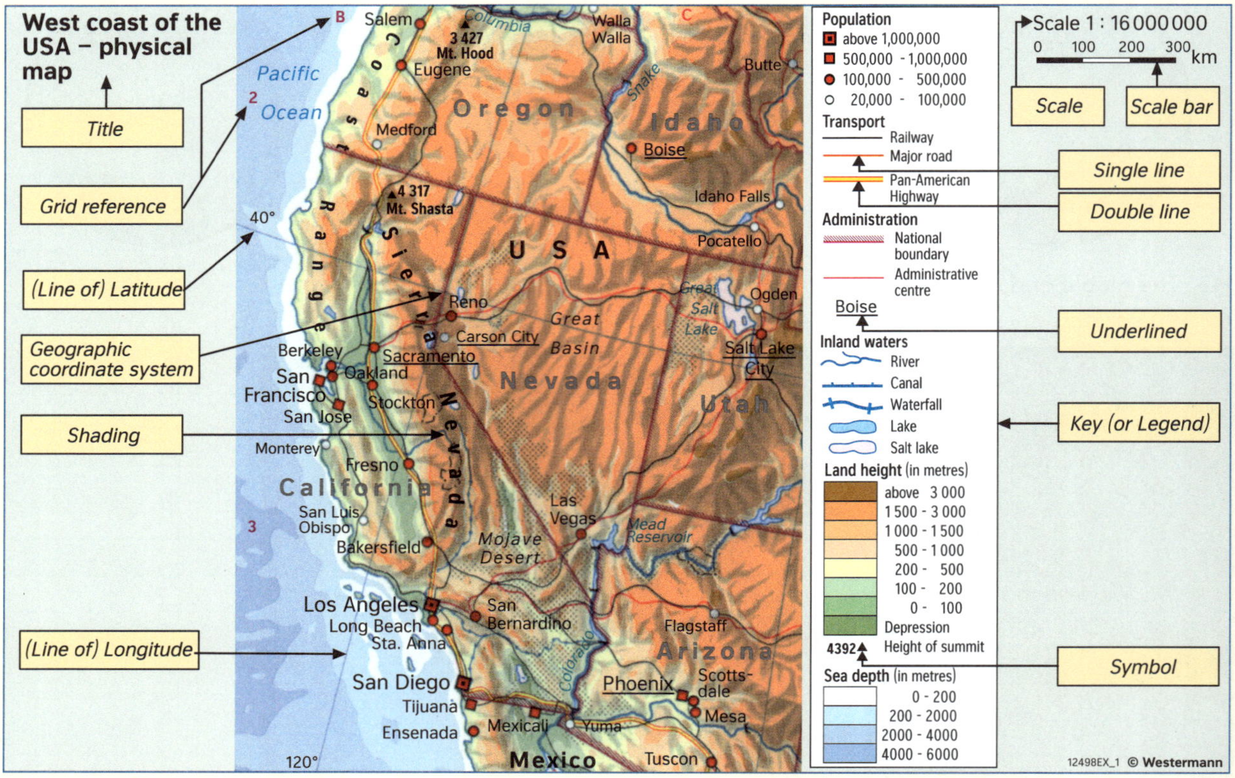

M1 Elements of a physical map

Elements of a physical map

Maps are the most important tools for geographers. Many tasks in Geography therefore involve the atlas. There you find many different maps which are either physical maps like **M1** or thematic maps (› p. 10).

The title of a map gives you a first orientation. The scale and the scale bar help you to measure distances. How to work out distances and sizes from a map see ›pp. 10-11.

Each map has got a key. You must study it carefully because it explains the colours, textures, lines and symbols on a map. Physical maps show the location of settlements, the relief (mountains and valleys), height of the land, depth of the sea, inland waters (rivers and lakes), transport routes, and political borders. A physical map is thus an important tool to get an understanding of a landscape.

Keep in mind

Maps do not show everything: they leave features out, they generalise and emphasise. In addition, like satellite images, they always show the past. Lakes can shrink, cities can become larger or smaller, roads can be built, and railway lines can disappear. When you describe a map, you have to take that into account.

If, according to the physical map, a city has a population between 100,000 and 500,000 and another city has between 500,000 and 1,000,000 inhabitants, we cannot say for sure if these cities are much different in size and rank. So, to be precise, we often have to use additional sources to find out more.

You describe a river always with the flow of its water, from the source to the mouth, from higher to lower regions.

100800-016 schueler.diercke.de

DE-015 www.diercke.com

GEO SKILLS

Describing facts in a physical map

Model text to describe central California in the map M1

Intro-duction

The **title of the map** I am referring to is 'West Coast of the USA – Physical Map' on the opposite page.
The map **has a scale** of 1 : 16 000 000.

Location

In my description **I will focus on** central California which is **the area around** Central Valley. **It lies in the west of** the USA, **bordered by** the Pacific Ocean.

Size

Central Valley **stretches over** 800 kilometres from north to south. Its **width** varies between 50 and 100 kilometres from west to east.

Relief

There are two high mountain ranges which **flank** Central Valley. T**o the west there is** the Coast Range which **runs parallel to** the coastline of the Pacific.
Its mountains **have an average height** of about 1,500 to 2,000 metres. East of the valley runs the Sierra Nevada.
Mt. Whitney is **the highest mountain**. Its **peak** is 4,421 metres high.

Bodies of water

The Sacramento River **coming from** the north and the San Joaquin River from the south **flow through** the valley. They **join** near San Francisco and **drain into** the Pacific Ocean.

Cities

San Francisco is **the largest** city in this area. **It is located at the coast** of the Pacific Ocean **at about** 38 degrees north and 122 degrees west. San Francisco **has a population of** 500,000 to 1 million inhabitants. Other big cities **nearby** are Berkeley **in the north-east** and Oakland **east of** San Francisco. Each of these cities has got about 100,000 to 500,000 inhabitants. San José, which has a similar size to San Francisco, **lies south-east of** it. Thus, San Francisco **is the centre of** a large conurbation.

Transport

A railway line **links** San Francisco **with** Los Angeles in the south. A major road, the Pan-American Highway, **follows** Central Valley and **connects** its main cities, such as Sacramento, Stockton, and Fresno.

Conclusion

Central California **is characterised by** two parallel mountain ranges, **separated by** a wide valley with two rivers. The main conurbation is around San Francisco, **mostly situated** in Central Valley.

HELPFUL WORDS AND PHRASES

to describe the content of maps:

Title – scale – border – region – area – located – situated – nearby – further away – relief – divide – connect (to) – mountain (range) – valley – steep – flat – undulating – body of water – river – source – mouth – tributary – lake – sea / ocean – settlement – city – population – inhabitants – transport – road – railway line – canal – pipeline – run (to) – lead (to).

HELPFUL WORDS AND PHRASES

to describe the location of a place:

A place is located / situated / lies

- in (a country, state, region, ...)
- (e.g.) east of ...
- on the northern/eastern/southern/western side of (e.g.) the mountains.
- in the north/east/south/west of ... (a country, state, region ...)
- in the centre of / at the edge of ...
- on a river / on the (e.g.) Pacific Ocean, ...
- near / close to / nearby a city / border.
- at about 38° N and 122° W.
- about 600 km south of ...

to describe transportation routes:

A canal / pipeline / road / railway line

- runs / leads from (e.g.) Fresno to Stockton.
- runs / leads along (e.g.) the coast.
- runs parallel to (e.g. a mountain range).
- follows the course of (e.g. a river).
- comes from (e.g.) the north / (e.g.) San Diego.
- connects / links ... (e.g. location A) with ... (e.g. location B).

A river

- flows from (e.g.) north to south ...
- flows through (e.g. a region, a valley) ... .
- joins ... (e.g. another river).
- drains into ... (e.g. a body of water).

to describe size / extent / height:

A region / An area

- is (e.g.) 20 km wide and 30 km long.
- has got a size of (e.g.) 200 km^2.
- covers an area of (e.g.) 50 km^2.
- stretches over (e.g.) 30 km from (e.g.) east to west.
- extends (for) (e.g.) 50 miles into (e.g.) the desert.
- extends as far as ...

The peak of ... (e.g. a mountain, a mountain range)

- is (e.g.) 4,421 m / up to (e.g.) 2,000 m high.
- reaches a height of (e.g.) 4,421 m.

The elevation of (e.g. a mountain or mountain range)

- is between (e.g.) 1,000 and 1,500 m.
- reaches (e.g.) 3,000 m.

Describe Thematic Maps

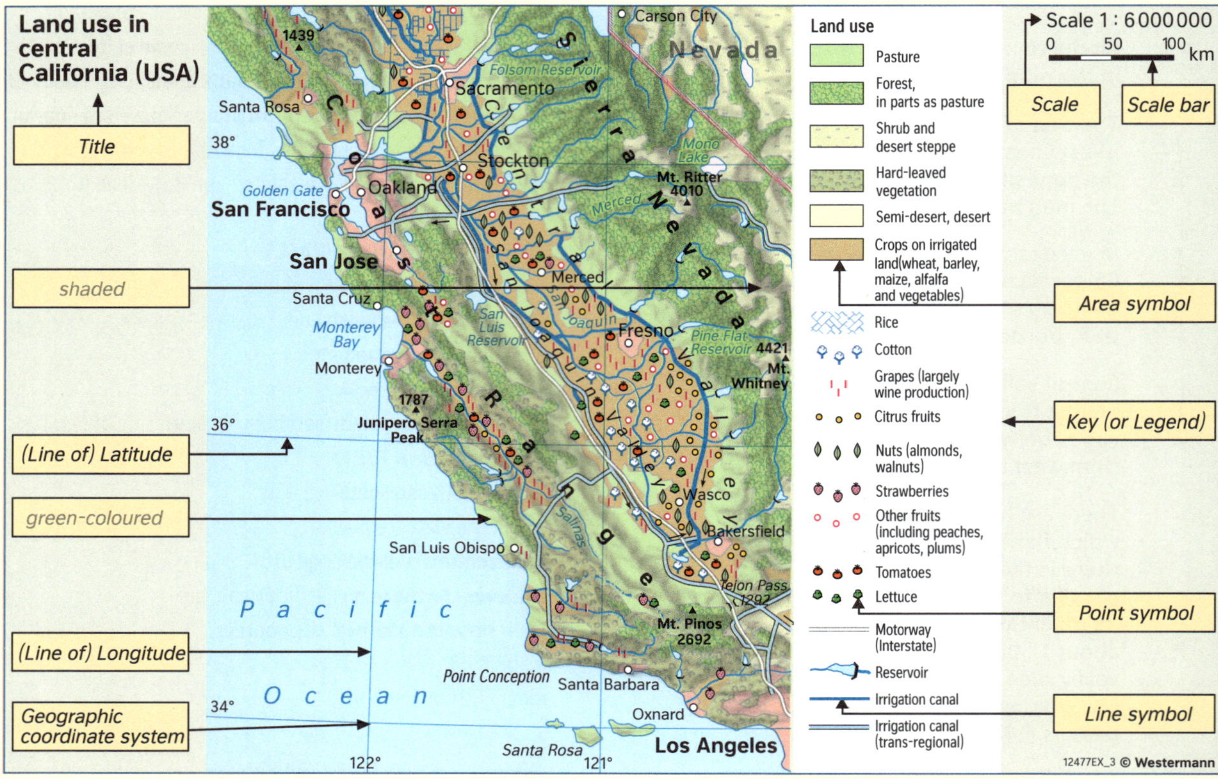

M1 Elements of a thematic map

Thematic maps are important tools for geographers. The most common thematic maps in your atlas deal with:

- nature (e.g. climate, vegetation, geology, etc.),
- agriculture (e.g. land use, types, crops, etc.),
- industry (secondary and tertiary industries),
- economy (e.g. world trade, exports and imports, etc.),
- environment (e.g. pollution, energy, etc.),
- population (e.g. density, growth, migration, etc.).

Most thematic maps use a great variety of symbols. Colours are used for areas. Different tones of shading and kinds of textures are used to give additional information. Lines represent connections and borders. Points may be abstract signs, as, for example, a symbol for the size of a city. However, point symbols may sometimes resemble reality. As an example, you can take the symbol for the chemical industry which resembles a flask.
Other examples are the symbols for crops or animals. The size or density of the point symbols indicate the amount of what the symbols stand for.

100800-012, -014
schueler.diercke.de

DE-016, -018, -020
www.diercke.com

How to describe a thematic map

When asked to describe a map you should firstly look at the title, which tells you the theme or main topic of the map . Then you should check the scale and the scale line, which give you an idea of the size of the area presented. The key or legend shows the symbols used on the map. Their arrangement and distribution contain important information for the geographer and therefore have to be described accurately.
The thematic map **M1** focuses on the agricultural activities in central California. When it is your task to describe the agricultural land use in San Joaquin Valley in California, the following questions should be considered:

- Which land use dominates San Joaquin Valley?
- Which crops are grown there, and to what extent?
- Where are they grown?
- What is the relationship between the location of the different crops, the climate, the relief, the course of rivers or canals, important transportation routes, or major settlements in this particular region?

SKILLS HACK

How to work out distances and sizes on a map

1. Measure the required distance on the map in cm.	Example in **M1**: the map's distance between Stockton and Bakersfield: 5.95 cm
2. Multiply this value with the denominator of the scale.	Scale 1/6,000,000. 5.95 x 6 000 000 = 35,700,000 cm
3. As the distance in reality is always given in km, you have to convert cm in km: 1 km = 100,000 cm.	35,700,000 / 100,000 = 357 km
4. You have the distance. Give rounded figures.	The distance between Stockton and Bakersfield is around 360 km
5. To work out the surface area you have to multiply length by width in square kilometres. Give rounded figures.	Surface area of San Joaquin Valley: Map: Length 6,5 cm, Width (av.) 1,2 cm. Reality: 390 km, 72 km. Area: 390 x 72 = (ca.) 2,800 km².

Model text to describe the land use in San Joaquin Valley, California (M1)

Intro-duction

The title of the map I am referring to is 'Land use in central California (USA)'. It can be found on the opposite page as M1. It has got a scale of 1 : 6 000 000.

Location and area

San Joaquin Valley in California **is located between** 35 and 39 **degrees** north and **stretches over a distance of** between 50 and 100 kilometres from west to east and around 390 kilometres from north to south. **It is bordered by** the Coast Ranges in the west and the Sierra Nevada in the east.

Topic general

The floor of San Joaquin Valley is **mainly used for** farming, **mostly** on irrigated farmland. Water is taken from San Joaquin River and its tributaries, often with reservoirs, which **lie** on the western slopes of the Sierra Nevada. **In addition to that**, canals carrying water not only from these rivers, **are used for** irrigation.

Topic specific

Between the towns of Stockton and Merced lettuce, tomatoes, fruits, and nuts **are grown**. Further to the south, in the centre of the valley, **we find** cotton, as well as citrus fruit growing on the valley floor. Vineyards **dominate** the eastern side of the valley, and stretch **along the foot of** the Sierra Nevada. **Only occasionally** the valley floor serves as pasture. The agricultural area of San Joaquin Valley is **well connected to** consumers in the big cities of Los Angeles and San Francisco by motorways, which link the major cities.

Summary

It can be concluded that the San Joaquin Valley is an area with an intensive and varied agriculture with subtropical and market gardening crops. The latter **have a good connection to** their markets.

HELPFUL WORDS AND PHRASES

to use when you describe a thematic map:

Title – scale – border – region – area – fragmented – compact – lie of the land – located – situated – nearby – further away – straight – curved – parallel – shape – size – connect (to) – transport – run (to) – lead (to) – distribution pattern – dispersed – dense – concentrated – regular – irregular – artificial – patchy – linear – intensive – extensive – in comparison

to describe location and distribution:

- The floor of the valley is used for ...
- Agriculture is dominated by ... in ...
- The production of ... concentrates north of ...
- ... is mainly grown / cultivated in ...
- ... stretch along ...
- ... are mainly located in ...
- ... are cultivated all over ...
- Farmers are concentrate on growing ... in (e.g.) the south.
- ... fields are scattered all over ...
- ... cluster near ...

to compare (for more › p. 36):

- The production of ...
  - is higher / lower than ...
  - has got a higher / lower value than ...
  - is of a higher / lower value than ...
  - shows a greater / smaller amount of ...
- Most of the north is used for ...
- ... is grown in the north to a great / small extent.
- The range of ... is wider in ...

to describe connections and relations:

- ... are closely / directly / loosely connected to ...
- Regions are linked with each other by ...
- The land use of this area is / can be / may be / is possibly related to ...
- The region has access to ...

Describe Photos and Other Pictures

M1 Positions in a photo

Most pictures in a geography book are photographs. Think of them as a window into a world which you do not know. They help you to understand the topic.
Describing them makes it possible for you to talk in class about the subject. The words in the photo **M1** help you to describe the positions in all photos and many other images as well. You find many other useful words in the switch boxes and in the model text. The photo is a view over a village of the Inuit in northern Canada.

Five top tips

1. Describe the picture systematically: go from left to right, from the foreground to the background, from the centre to the outside, clockwise or anticlockwise. You may start the description with the most important detail, too.
2. For every detail you mention note its position in the photo.
3. Describe what you can really see. If you are not sure, make an educated guess (› p. 35).
4. Use many words and phrases from these pages for a varied description.
5. Use technical terms from the topic.

Five helpful questions

1. When and where was the photo taken?
2. What kind of people / activities / landscapes are shown?
3. How does the photo help me to understand the topic?
4. How does the situation in the photo differ from where I live?
5. What is the effect of the photo on me?

Model text to describe a photo of an Inuit village (M1)

Introduction

The photo **M1** on the opposite page **was taken at an oblique angle** on a sunny day. **The caption states that** it shows an Inuit settlement in springtime.
Nearly everywhere there is snow and ice.

Description

On the right hand side of the photo and **in the background**, the sea is covered with ice. **You can see** several icebergs **in the back** and pack ice **along** the coast. **Beyond** the ice there are many boats on the shore to be seen **in the right top hand side of the photo. On the left hand side of the photo, a little away** from the shore, are the houses in which the Inuit live. Like all the other buildings and boxes you can see they seem to be made of wood. The hut **below the centre of the photo** looks derelict. **Near** the houses in **the centre** you can see clothes on washing lines. The only person you can see **in the photo** is **in the foreground**. He (or she) is **next to** several dogs, some of which are bound together by leashes. The wooden structures **in the bottom right hand corner of the photo** appear to be for the dogs, although the biggest one **nearest to the centre does not seem to have a** roof.

Educated guess

I think the photo was taken in late spring because the huts cast a shadow, so the sun is quite high in the sky. **Another hint** is that the ice is still there.

Note: *There are several other types of images, for example: (Oil) painting – figure – cartoon – sketch – illustration – (technical) drawing – (block) diagram – collage – poster. The skills to describe these pictures are the same as they are for photos!*

HELPFUL WORDS AND PHRASES
to describe a photograph:

A bird's-eye view A photograph An aerial (photo) The close-up	• gives you information about ... • draws the attention to ... • illustrates ... • shows ... • tells me ... • deals with ... • focuses on ... • was taken ...
The caption says / states that ...	

HELPFUL WORDS AND PHRASES
to describe positions:

Something ... • can be found • is located / situated / positioned • lies	• above / atop ... • at some distance. • at the very edge. • behind ... • below ... • beyond ... • on the far left / far right. • further away (from) ... • in front of ... • nearby / close by • next to ... • on the right / left / other side. • top left / right.

HELPFUL WORDS
to describe objects:

The words below give you some ideas to make the description of pictures more varied and interesting.

Shape
round – curved – pointed – blunt – globular – spherical – stretched (out) – cylindrical – cubical – rectangular – squared – triangular.

Size
long – tall – great – wide – broad – large – huge – big – short – small – little – minute – tiny.

Colour
bright (red, ...) – light (green, ...) – dark (blue, ...) – (yellow, ...) ish – off-(white, ...).

Pattern
rough – smooth – uniform – varied – regular – irregular – dotted – plain.

HELPFUL WORDS AND PHRASES
to use when you are not quite sure:

I suppose ...
It seems / appears to be ...
I think it is ...
... as if it were ...
(To me) it looks like ...
(Do not forget to give reasons for your suggestions!)

How to Work with Aerial Photographs

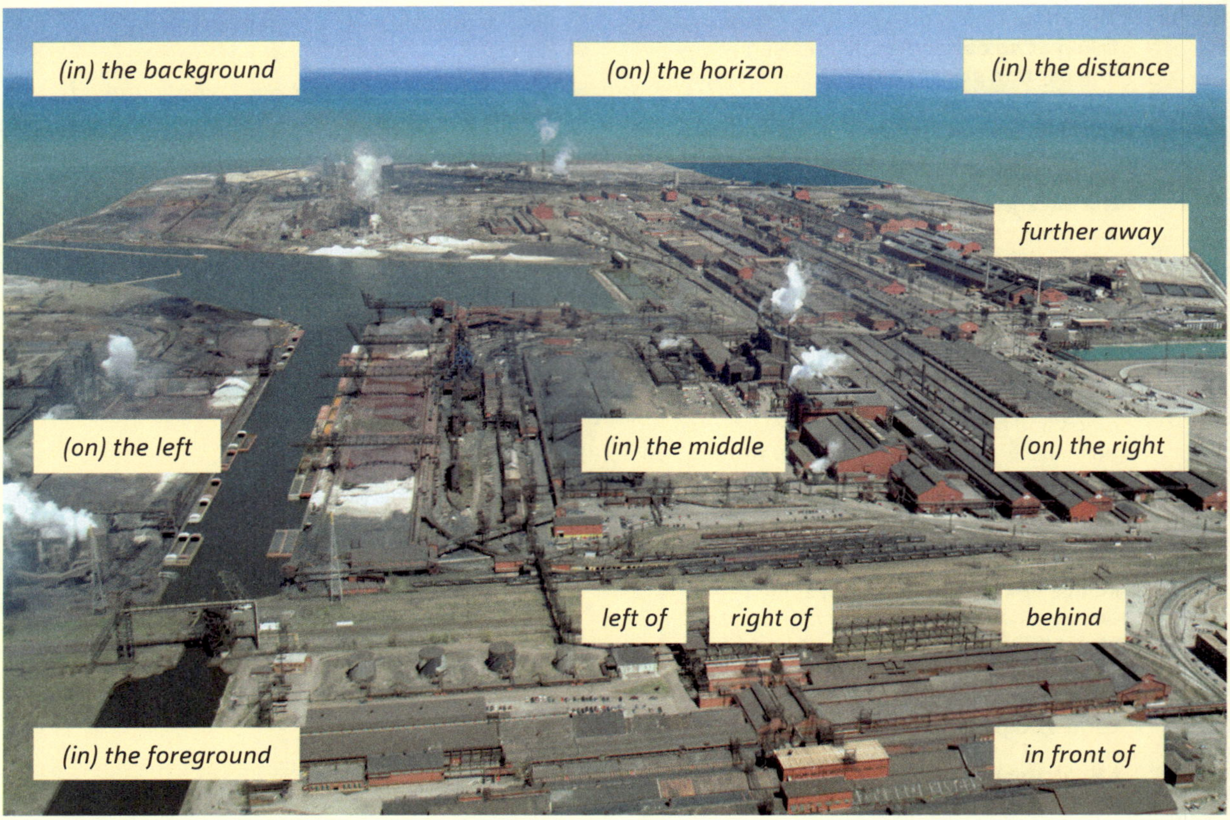

M1 Positions in an oblique aerial photo

Having information about the Earth's surface is big business today. Satellites and aeroplanes deliver images which help us to forecast the weather and provide us with information about mineral resources, environmental changes, next year's harvests, and even military operations. They use the visible light spectrum as well as infrared and radio waves in a method called remote sensing. The pictures they produce are called images.

Aerial photos use visible light. They are all photos taken from above, either directly overhead or at an oblique angle. Vertical aerial photos look directly down on an area the same as a map does. Oblique aerial photographs, like **M1**, look down on an area at an angle (**M2**), so they have fore- and background.

http://www.diercke.com/diercke-globe/download.xtp
http://www.google.co.uk/intl/en_uk/earth/

100800-012 schueler.diercke.de

DE-008 www.diercke.com

Vertical aerial photos

There are many websites on the internet which show aerial photos, taken by aeroplanes or satellites, shot from directly above (**M3**). Often you can zoom in and see the landscape in great detail, sometimes in 3D. You can also zoom out to find out, which area you look at. When you have a photo, often the location in the global grid is given. With an atlas or the internet you can find out the exact location.

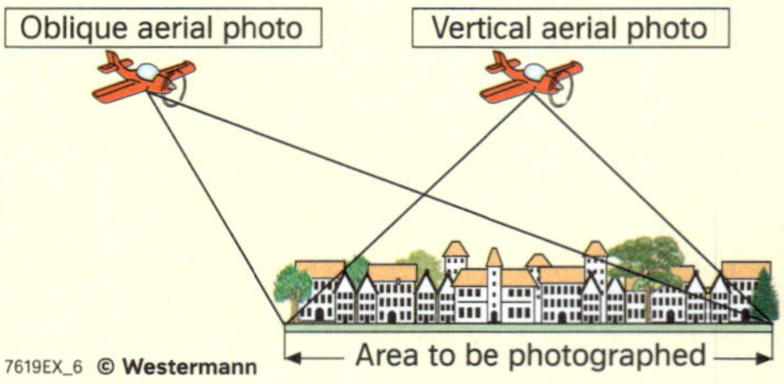

M2 Types of aerial photographs

Model text to describe an oblique aerial photograph of an industrial area (M1)

Intro-duction

The oblique aerial photo M1 on the previous page shows an industrial area in the Manufacturing Belt on the Great Lakes.

Description

The photo **was taken at an oblique angle** from an aeroplane on a sunny day. **Nearly everywhere there is** smoke which is coming out of many chimneys.
On the left of the aerial photo there is a canal, **running from the foreground to a** harbour basin **further away**. From the basin a wider canal **leads to** the open sea which **can be seen in the distance**. There are many factories **along the** canal which is **crossed by** a bridge **in the foreground**. Some factories **are connected by** conveyor belts. The factories often have heaps of some sorts **nearby**. These are transported by ships which dock **alongside** the quays **in the middle** of the aerial photo. There are green strips of land alongside railway tracks and roads which **cross** the canal **at a right angle in the foreground** of the photo **behind** two factories.
In front of the railway line there is a row of storage tanks.

Sum-mary

All in all it looks like an old industrial area which uses or / and produces bulk material, **probably** in the metal producing industry.

Model text to describe a vertical aerial photograph of an industrial area (M3)

Intro-duction

In M3 you see a satellite image taken from around 450 m altitude.
North is at the top.

Loca-tion

The position in the global grid indicates that it is located at 40°42' N **and** 73°59' W, which is in Brooklyn in New York City.

Description

The area has easy access to the East River which is in the east. **You can make out** two long chimneys on a building **in the centre** with round oil tanks **east of** them. A coal bunker is **nearby**. Coal bunker and oil tanks have quay sides for ships.

Conclu-sion

The buildings and the location **suggest that the aerial shows a** New York thermal power station.

HELPFUL WORDS AND PHRASES
to describe aerial photos and satellite images:
The oblique / The vertical aerial photo(-graph) / aerial shows ...
The Landsat / The satellite image reveals ...
The photograph was taken looking north / east / south / west / across / over / towards ...
Remember: The words to describe pictures in general (› pp. 12-13) are useful to describe aerial photos, too!

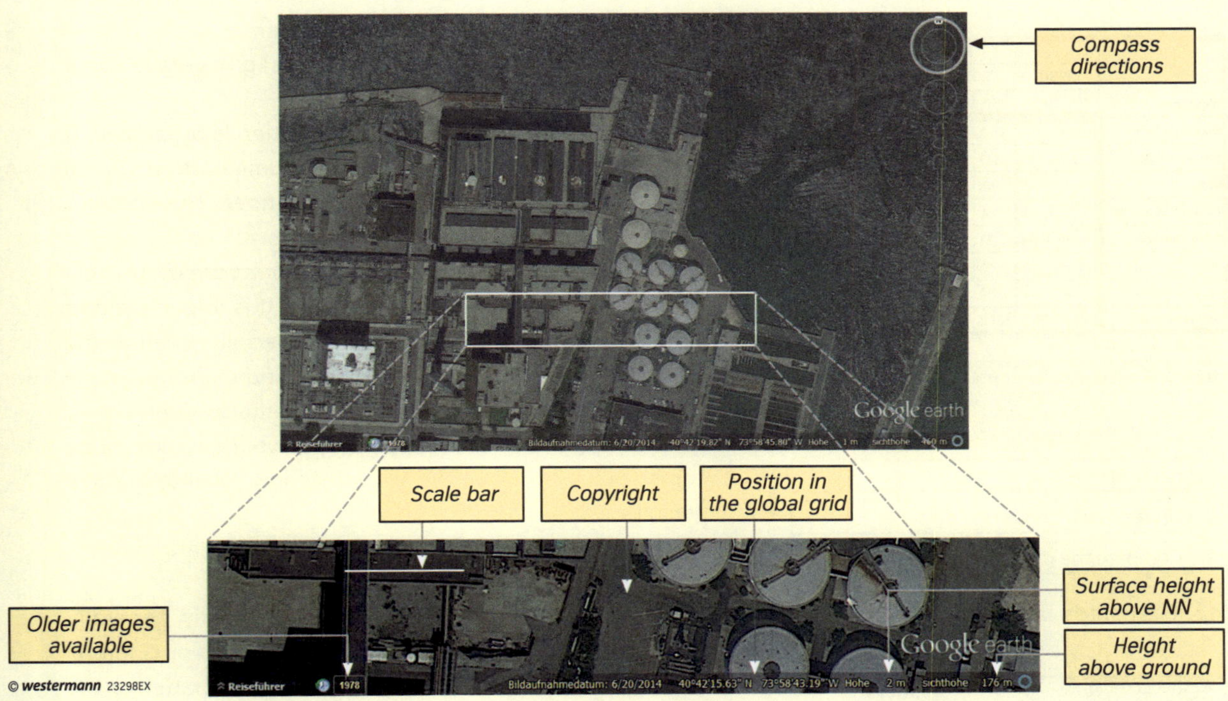

M3 Elements in a satellite image

Comprehend and Describe Tables

Column | Column head | Absolute figures | Relative figures | Box head | Row head | Row | Data field | Cell | Table footer | Caption (title) | Source

Ethnic group	Population							
	1990		2010		2030		2050	
	Million	Per cent	Million	Per cent	Million	Per cent	Million	Per cent
White, non Hispanic	188.1	75.6	196.8	63.8	198.8	55.5	186.3	46.6
Hispanic (any group)	22.4	9.0	50.5	16.4	78.7	22.0	111.7	27.9
Black	30.0	12.1	38.9	12.6	45.5	12.7	52.0	13.0
Asian & Pacific Islanders	7.3	2.9	10.6	3.4	22.0	6.1	29.6	7.4
Native Americans	2.0	0.8	2.7	0.9	2.7	0.8	2.9	0.7
Other group	9.8	3.9	19.1	6.2	n. a.	n. a.	n. a.	n. a.
Two or more groups	n. a.	n. a.	9.0	2.9	10.1	2.8	16.4	4.1
Total population	**248.7**	**100.0**	**308.7**	**100.0**	**358.5**	**100.0**	**399.8**	**100.0**

n. a. = Data not available — Source: www.census.gov

M3 Population development in the USA 1990-2050 (2030 and 2050 are estimated)

M1 Elements of a table

A table is a grid system with information in the form of text and numbers. It is used to present information in an ordered way and helps to find information quickly.

Ethnic group	Population							
	1990		2010		2030		2050	
	Million	Per cent	Million	Per cent	Million	Per cent	Million	Per cent
White, non Hispanic	188.1	75.6	196.8	63.8	198.8	55.5	186.3	46.6
Hispanic (any grop)	22.4	9.0	50.5	16.4	78.7	22.0	111.7	27.9
Black	30.0	12.1	38.9	12.6	45.5	12.7	52.0	13.0
Asian & Pacific Islanders	7.3	2.9	10.6	3.4	22.0	6.1	29.6	7.4
Native Americans	2.0	0.8	2.7	0.9	2.7	0.8	2.9	0.7
Other grop	9.8	3.9	19.1	6.2	n. a.	n. a.	n. a.	n. a.
Two or more grops	n. a.	n. a.	9.0	2.9	10.1	2.8	16.4	4.1
Total population	**248.7**	**100.0**	**308.7**	**100.0**	**358.5**	**100.0**	**399.8**	**100.0**

n. a. = Data not available — Source: www.census.gov

M3 Population development in the USA 1990-2050 (2030 and 2050 are estimated)

1. Check the title and source
2. Scan the heads
3. Understand the units
4. Compare the data by rows and columns

M2 First steps in reading a table

100800-210 schueler.diercke.de

DE-136 www.diercke.com

How to read a table

STEPS

1. Identify the title, so that you know what it is all about. Find out the source to see if the data are reliable. Data from official sites are normally in order.
2. The row headings (stubhead and stubs) give details about the data.
   The boxhead shows how the information is organized. It contains all the information for the columns. Often several columns share one head, called spanner. The spanner applies to all the columns it covers.
3. Somewhere in the column heads (or sometimes in the row heads) are the units of the data field. It is important that you understand the units clearly, especially whether the data are absolute figures or relative figures (for example percentages). When the unit is a percentage, determine if it adds up to 100 per cent by going down the column or by going across the row. This will affect how you interpret the results.
   Relative figures are often used to compare data of different categories or sizes.
4. It is important to read the table systematically, generally from left to right or top to bottom. The rows or columns are often ordered according to time, the alphabet, or the numbers from the highest to the lowest values or vice versa (**M2**).

SKILLS HACK

How to describe a table

When you describe the content of a table, it should be systematical and in tight focus of the task given. Tables contain so much information, most of the time much more information than the one you have to address in your task.

Often tables are not sorted according to the values. So it can be a good idea to rank the data from highest to lowest value.

How to turn absolute figures into percentages

Question: How many per cent of the population of the USA in 2010 were Black?

Calculation: (Number you want to turn into per cent) / (total) * 100 = (Percentage you are looking for).

Here: 38.9 / 308.7 * 100 = 12.60123 % ≈ 12.6 %.

You say: The percentage of the black population in 2010 is 38 point 9 million divided by the total population in 2010 which is 308 point 7 million, times one hundred equals 12 point 6 0 1 2 3 per cent, rounded 12 point 6 per cent.

Answer: 12.6 per cent of the population of the USA in 2010 were Black.

(More on the subject absolute figures-percentages ›pp. 20f, ›p. 27)

Model text to describe a table about the development of the Black population in the USA between 1990 and 2050 (M1)

Introduction

The table 'Population development in the USA 1990 to 2050' **is based on data from** the website *www.census.gov*.
The data from 2030 and 2050 are **predictions**. They **are given in absolute figures** and **as percentages**. They **are divided between** five different ethnic groups and one further entry, 'other race'. My **task was to focus on** the black ethnic group.

Description

Between 1990 **and** 2050 the black population **is estimated to increase** from 30 million to 52 million people. Although the number of Blacks **presumably** will increase **by** 75 **per cent** over these sixty years, the percentage of the Blacks within the projected US population will **only grow slightly** from 12.1 per cent to 13.0 per cent. This is **due to the fact** that the total population **is supposed to** increase from around 250 million in 1990 to nearly 400 million in 2050. In 1990, Blacks were **the second largest** ethnic group, whereas in 2050 they will be the third largest ethnic group. The **biggest per cent increase** of Blacks **in a** twenty-year **period** was **between** 1990 **and** 2010, when **the share of** the black population **rose from** 12.1 **to** 12.6 per cent.

Summary

Between 1990 and 2050 **around one** in eight Americans belongs to the black ethnic group.

HELPFUL WORDS AND PHRASES

to describe tables:

This table is about ... (title). Its source is ... (source).
I focus(ed) on ... (task).
My task is / was ...
According to the data ...
The data show / indicate ...
Further down the column / across the row it says that ... .
This means that ...
The total does (not) add up to one hundred per cent. It is instead ... per cent.
It / The reason could be that ...

to round off figures:

Approximately – around – about – roughly – rounded off – circa – almost – nearly – slightly more / less than ... – little more / less than ...

to describe changes:

Upward change: grow – rise – climb – increase.
No or little change: stay the same – continue – show (nearly) zero growth.
Downward change: sink – fall – drop – decrease.
Regular up and down change: fluctuate / vary between ... and ... .

Note: *There are more words and phrases to describe changing figures on the following pages!*

Draw and Describe Line Graphs

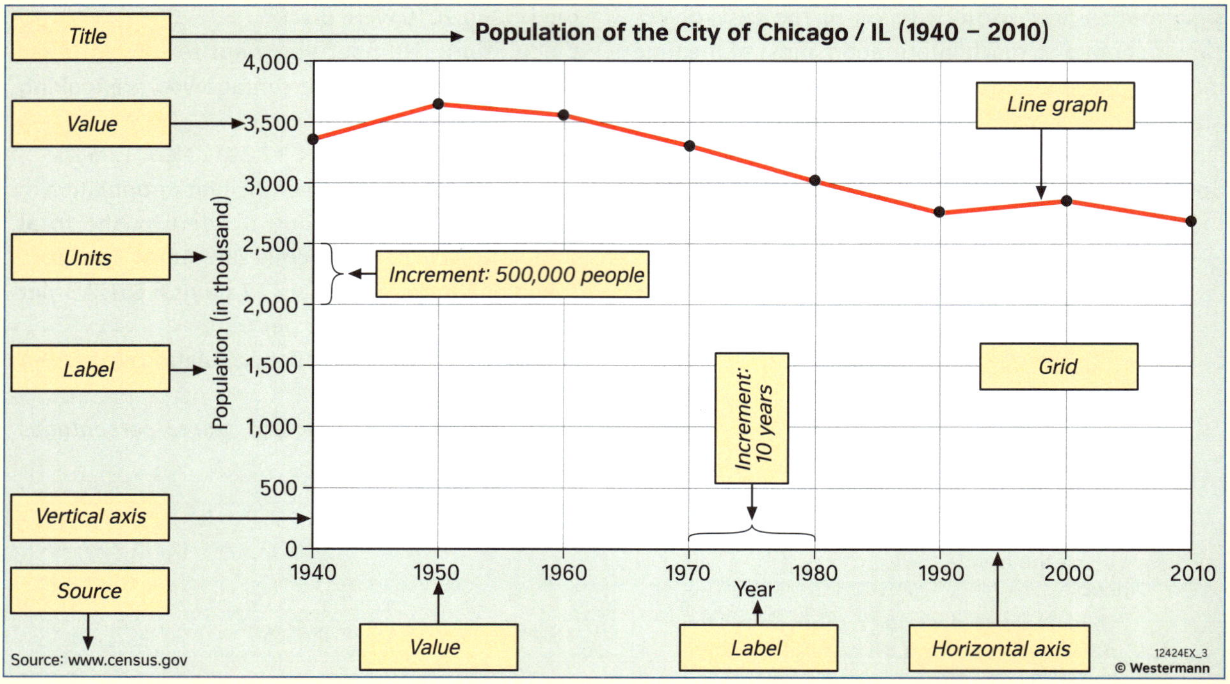

M1 Elements of a line graph

Graphs help us to visualize data. So you use graphs frequently when the data are too complex to understand them quickly. The graphs which geographers use most are line graphs, bar charts, and pie charts.

Year	Population (Million)
1940	3.40
1950	3.62
1960	3.55
1970	3.37
1980	3.00
1990	2.78
2000	2.90
2010	2.70

M2 Data table

Line graphs and bar charts are diagrams with a horizontal and a vertical axis which form a grid. This helps to read the data more easily compared to a table (**M2**). Line graphs are often used when data are plotted for a certain period of time.

Note: *The figures only show the data of that particular year, for example 1950 = 3.62. It can well be that the maximum of the population development was in another year which is not shown in the data.*

DE-025 www.diercke.com

How to draw a line graph

STEPS

1. Use a ruler to draw straight lines.
2. Assign the horizontal axis to the independent variable (cause), and the vertical axis to the dependent variable (effect).
3. Check the maximum and minimum data to find suitable units.
4. Look at the data to check how much space you need. For more clarity, use all the space available.
5. Make sure that the data range available for the graph is a little larger than the data you have to plot.
6. Use increments on both axes which are regularly spaced and mark the values.
7. Label both axes.
8. If necessary, write the units behind the label in lowercase and in brackets.
9. Find a title to inform the reader what the graph shows.
10. Centre the title at the top.
11. Add a key if more than one set of data is in the diagram. Then use different colours for clarity.
12. Plot the data clearly and correctly and connect them with lines (ruler!).

Keep in mind

Before you start the description, divide the graph into several parts; but as few as possible. To do so, look for parts of the graph which show similar characteristics. In **M1**, you can divide the graph into four sections: from 1940 to 1950, from 1950 to 1990, from 1990 to 2000, and from 2000 to 2010.

Note: *Do not describe a graph, describe the facts which the graph represents!*

Model text to describe a line graph about the population development in Chicago (M1)

Introduction

The line graph shows the population development in the City of Chicago / IL between 1940 and 2010. **It is based on data from** the USA government.

Description

During the first ten years the number of people in Chicago **grew only slightly** to over 3.6 million inhabitants in 1950, when the city's population **climbed to a maximum**. From 1950 onwards there **was a continuous and steady decline to** around 2.8 million city dwellers in 1990, when **a minimum was reached**. At the end of the next ten years, **a slight increase** in Chicago's urban population was observed. However, the last ten years saw **a further decline in** population size to around 2.7 million inhabitants in 2010.

Summary

Over the seventy years from 1940 to 2010 the city of Chicago **lost more than** 700,000 inhabitants.

HELPFUL WORDS AND PHRASES
to describe line graphs (and changes in value in other diagrams and tables):

• The diagram • The line graph	• provides • offers • gives	information about ...
	• is based on material by ... • uses data from ...	
It shows the	• (title)	
It ... • represents the • shows the • indicates the • illustrates the	• development of ... • increase in ... • decline of ...	• ... between ... and ... • from ... to ...
Between ... / During ... / From ... onwards ... / Since ... / Over (the next) ... years ... / The following ... years ...		
• The number of ... • The amount of ... • The ... • The percentage of ...	• grew • rose • fell • declined • decreased • increased • dropped	• steadily • gradually • constantly • rapidly • slowly • slightly • sharply • moderately
The ...	• came to a standstill • remained constant / stable • levelled off • reached a peak / maximum • reached a low / minimum	

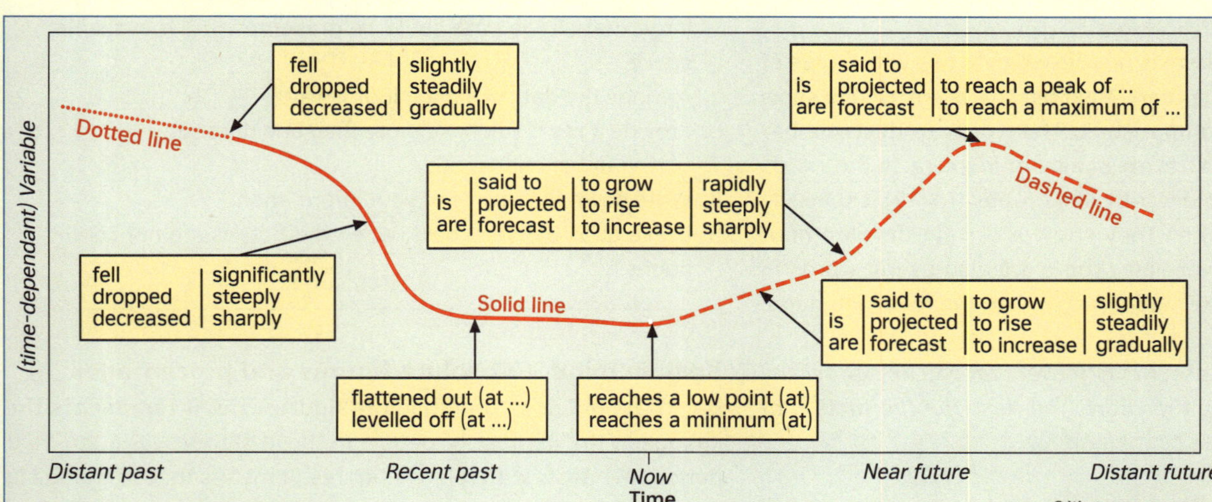

M3 Describing time in a line graph

Draw and Describe Bar Charts

Seven fastest growing states (population change 2000-2010)	
in per cent	
State	Per cent
Nevada	35.1
Arizona	24.6
Utah	23.8
Idaho	21.1
Texas	20.6
North Carolina	18.5
Georgia	18.3
by number	
State	Inhabitants
Texas	4,293,741
California	3,382,308
Florida	2,818,932
Georgia	1,501,200
North Carolina	1,486,170
Arizona	1,261,385
Virginia	922,509

Source: www.census.gov

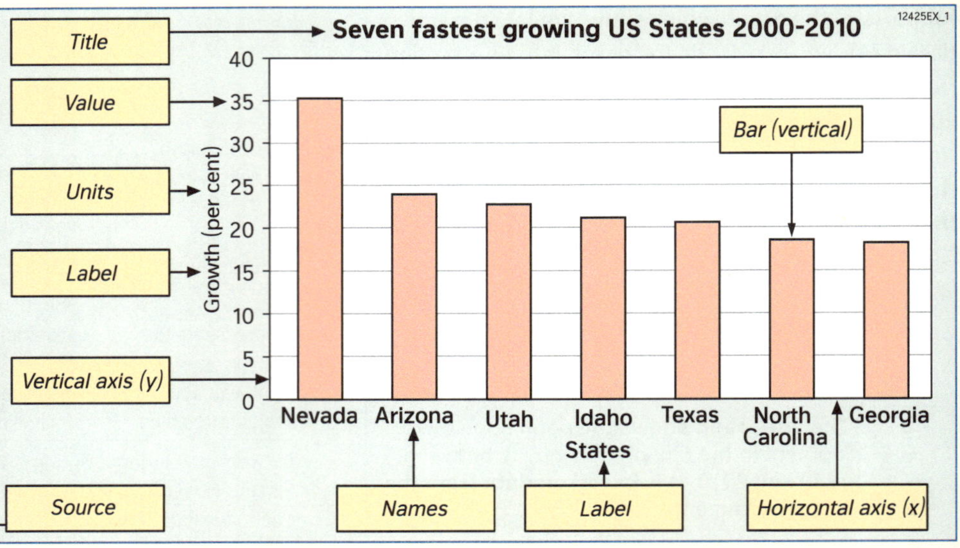

M1 Elements of a vertical bar chart (created from the data in the table left)

Bar charts (bar graphs) are a way to illustrate data. Instead of dots (as in a line graph, › p. 18) a bar chart uses a series of bars for the display.
There are two basic types of bar charts shown here. In vertical bar charts the bars run vertically (**M1**) whereas in horizontal bar charts the bars run horizontally (**M2**).
The vertical bar chart **M1** and the horizontal bar chart **M2** have been created using the data from the table left of **M1**.
In **M1** the seven US states are put on the horizontal axis, the growth is put on the vertical axis.
There is no rule when to use which type of bar chart. Most bar charts are used to show comparisons. The words to describe bar charts are similar or identical to the words to describe line graphs (› p. 18), especially when they show a certain development over time. There are also useful words in the model text and on the following page.

Note: *Never forget that you do not describe the chart. You describe the facts that the bars stand for.*

DE-025
www.diercke.com

How to draw a bar chart

STEPS

1. It is best to present your data in a bar chart, when the data for one axis are not (continuous) numbers.
2. Decide which data you put on the horizontal axis and which on the vertical one. (In a vertical bar chart, you usually place the categories being studied on the horizontal axis and the measurements, amounts, or percentages on the vertical axis.)
3. Put labels on the axes.
4. Decide on the scale – that means how much each unit on the axis is. The chart should always be as large as possible, so large that all the data could be presented in the available space.
5. Arrange the data in a meaningful way.
6. Provide a key if you have more than one data set.
7. Give your bar chart a title.
8. Write down the source of your information.
9. Leave gaps between bars, when the figures are not continuous.

Keep in mind – Absolute figures and percentages

Bar chart **M2** uses the absolute figures, i.e. it represents the change in the number of people in these states over a ten-year period. **M1** deals with relative figures. It means that th change in population size is expressed as percentage, based on the figures for the year 2000. **M1** and **M2** come from the same official US

GEO SKILLS

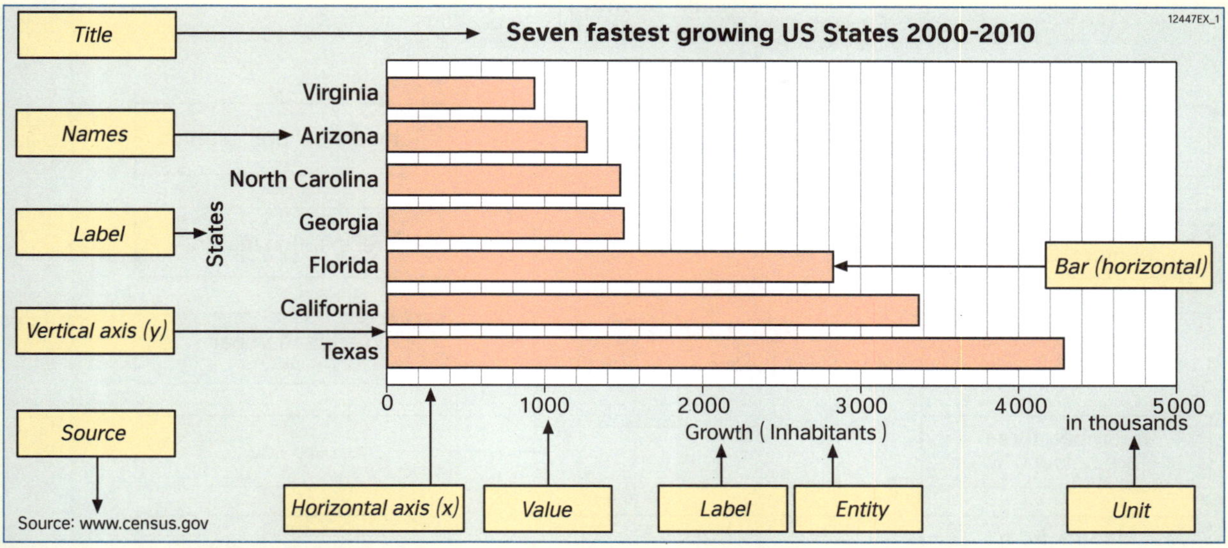

M2 Elements of a horizontal bar chart (created from the data in the table in **M1**, left)

Model text to describe a bar chart about the population growth in US States (M1)

Introduction

The bar chart shows the seven states in the USA with **the highest population** growth between 2000 and 2010. The states **are arranged in order from the highest to the lowest** growth in per cent of the population. These ten-year **growth rates reflect** the relative population increase in these states, **based on** the population of the year 2000.

Description

The chart shows that Nevada had the highest growth. **More than one third** of the number of people who were in Nevada in 2010 had not been there in 2000. The state with **the second highest** ten-year growth rate in the USA was Arizona, where the population size grew by **nearly a quarter** over these ten years. Utah was **in third place** with a **just marginally smaller** growth than Arizona. Idaho and Texas in fourth and fifth place had a population growth of **more than one fifth** during this period. The population of **sixth-positioned** North Carolina and Georgia in **seventh place** still grew more than 18 per cent between 2000 and 2010.

Conclusion

The states with the highest growth rates lie in the south and west of the USA. Their population grew by more than **one sixth** within ten years.

government statistic. But you can see that the states and the order of the states in **M1** and **M2** are totally different. For example, Nevada, the fastest growing state according to **M1**, is nowhere to be found in **M2**, whereas California, not in the chart in **M1**, is the fastest growing state according to **M2**.
Nevada's population size (2.7 million inhabitants) is much smaller than that of California (37.3 million inhabitants). So a growth of 35.1 per cent in Nevada means 702,295 people. The population increase of 3.4 million people in California between 2000 and 2010 meant an increase of 'only' 9.98 per cent.

So you cannot just say: 'Nevada was the fastest growing state between 2000 and 2010.' You have to say: *'Nevada was the state with the highest population growth rate between 2000 and 2010'* instead.

HELPFUL WORDS AND PHRASES
to describe relationships of data:

- (not / much / not much) larger than ...
- (not / nearly) as big as ...
- a little more / less than ...
- (just) over / above / under / below ...
- between ... and ...
- from ... to ...
- one in five = one fifth = 20 per cent
- one in four = one quarter = 25 per cent (etc.)

(More on the subject: › p. 26)

Draw and Describe Climate Graphs

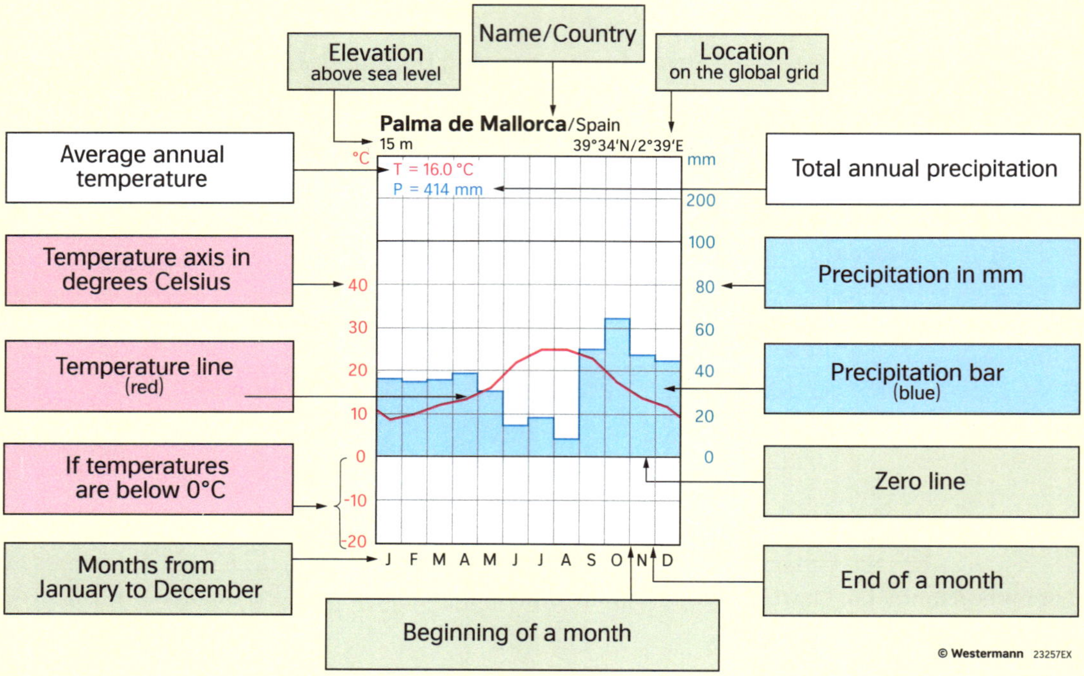

M1 Elements of a climate graph

How to draw a climate graph

A climate graph (**M1**) cannot forecast the weather, but it helps to describe and analyse the average climatic situation of a place (**M2**).

The drawing of a climate graphs follows fixed rules, which makes it easier to compare the climate of different places.

STEPS

1. Take a sheet of squared paper. Draw the base line and divide it into twelve months (1 month = 1 cm).
2. Draw the left axis for the temperature data (10 °C = 1 cm). If necessary, leave room for minus temperatures at the bottom.
3. Draw the right axis for the precipitation data (20 mm = 1 cm). When there are months with precipitation over 100 mm label the axis 20, 40, 60, 80, 100, 200, 300 mm and so on.
4. Draw the blue bars of precipitation for each month. A bar should fill the whole month in width.
5. Then make little dots or crosses in the middle of each month for the temperature and draw a continuous red line from one end to the other, connecting the dots.
6. Write down the name of the climate station, the elevation, and the location in the grid system above the climate graph.
7. Add the average temperature of the year and the average annual precipitation at the top left of the grid.

Month	J	F	M	A	M	J	J	A	S	O	N	D	Year
t (°C)	9	10	11	12	16	21	24	24	22	18	13	11	16
p (mm)	37	35	36	39	30	14	9	20	50	64	47	44	414

M2 Climate data of Palma de Mallorca / Spain, 8 m above sea level, 39° 34' N, 2° 39' E

How to describe a climate

STEPS

1. Start with the place name and country, the location on the global grid, and the elevation.
2. Name the month(s) with the highest and lowest temperature. Calculate the temperature range, i.e. the difference between the two.
3. Find the driest month(s) and the month(s) with the highest precipitation.
4. Characterise the seasons (see below).
5. Finish with the average temperature of the year and the average annual precipitation.

Model text to describe the climate of Palma de Mallorca (M1)

Introduction

I want to describe the climate of Palma de Mallorca in Spain, **located at** 39° 34' N and 2° 39' E. **Its elevation is** 8 metres **above sea level**.

Description

The coldest month is January **with an average of** 9 °C, whereas July and August **are the warmest months, averaging** 24 °C. Thus **the temperature range is** 15 °C. July is **the month with the least precipitation** (9 mm) and October **the wettest month** with 64 mm precipitation **on average**.

Summary

Over all, the **winters in** Palma de Mallorca **are mild and wet**, the **summers are warm and dry**. The **yearly average temperature is** 16 °C, whereas **the precipitation adds up to** 414 mm **per year on average**.

Keep in mind – Seasons

It is a good idea to look for seasons in the climate graph. In Palma, for example, there are four seasons due to changing solar radiation during the year. In the Northern Hemisphere winter is from December to February, spring from March to May, summer from June to August, and finally autumn from September to November.
In the Southern Hemisphere the seasons are as follows: Winter: June – August, spring: September – November, summer: December – February, autumn: March – May.
When the precipitation changes considerably over the year, you often separate a dry season (with little or no precipitation) from a wet season (with much precipitation).

100800-244, -246, -254
schueler.diercke.de

DE-025, -176
www.diercke.com

HELPFUL WORDS AND PHRASES

to describe the climate in general:
hot, warm, mild, moderate, cool, cold, humid, arid, dry, wet, rainy, high, low, continental, maritime.

to describe data from climate graphs:

Location
The climate graph shows the climate of ... (city) in ... (country).
Its location on the global grid is ... / It is located at ...
... lies ... m above sea level.

Temperature
The maximum / minimum temperature is ... The warmest / coldest month(s) is / are ... ... followed by ...
The temperature range is from ...° C to ...° C There are high / low temperatures in / during ...

Precipitation
The maximum / minimum precipitation is ... The wettest / driest month(s) is / are ... There is little / much precipitation in / during ... In / During ... the rainfall is high / low.

Seasons
The rainy season / dry season is ... from ... to / between ... and ...
In winter (etc.) time ... / During the winter (etc.) ...

Average
The average annual temperature / precipitation is ...
On average ... / ... on average.

Describe and Evaluate Age-and-Gender Diagrams

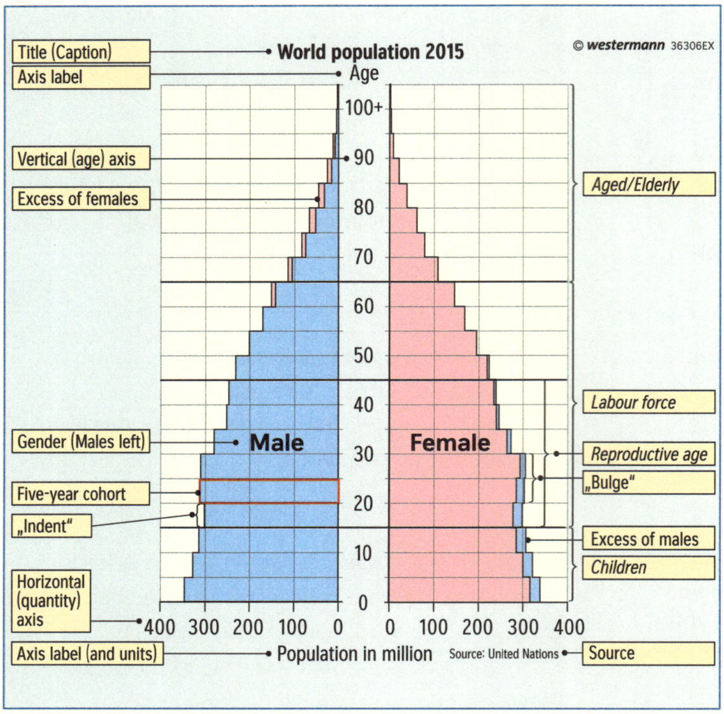

M1 Elements of an age-and-gender diagram

In an age-and-gender diagram demographic data are presented as two back-to-back horizontal histograms (= bar charts with no gap between bars), one for males (on the left) and one for females (on the right). The vertical axis shows age in five-year cohorts (five-year groupings), from the youngest (0 – 4 years) up to the oldest. The horizontal axis shows either the population of the cohorts in millions or in per cent of the total population. Thus, each horizontal bar represents one age-gender cohort, for example, males between 20 and 24 years of age (**M1**).

Age-and-gender diagrams graphically display long-term trends in the birth and death rates, but also reflect shorter term baby-booms, wars, migrations, and epidemics which appear as bulges or indents in the diagram, as these incidents mostly affect certain age groups.

How to evaluate an age-and-gender diagram

STEPS

1. Identify the overall shape of the diagram (**M2**).
2. Apply the characteristics from **M2** to your diagram.
3. Look at indents and bulges in the diagram. Use your background knowledge for explanations.
4. Give an outlook for future development of the population as an educated guess (›p. 35).

SKILLS HACK

Comparing age-and-gender diagrams

It is easier to compare the age structure of populations with each other when the cohorts are displayed as per cent or per mil of the total population.

This way you can compare populations of different sizes, like, for example, China and Germany or the world population at different times (**M3**).

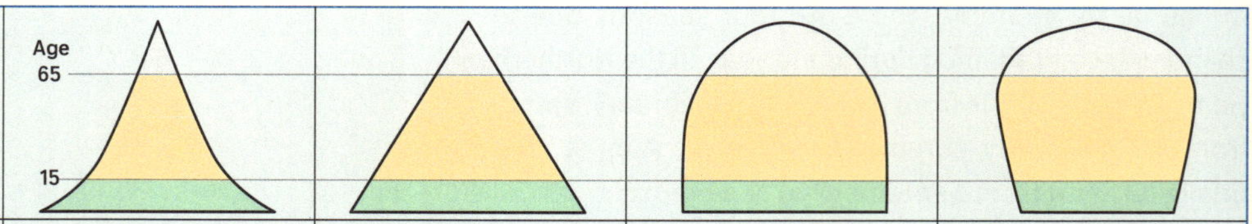

	Pagoda	Pyramid	Bell	Urn (Onion)
Birth rate	high	high	medium	low
Death rate	(very) high mortality	medium mortality	low mortality	increasing mortality
Life expectancy	very low	increasing	medium to high	high
Population growth	growing	fast growing	stationary	shrinking
Dependency ratio	low/medium	low	medium	high

M2 Basic characteristics of age-and-gender diagrams

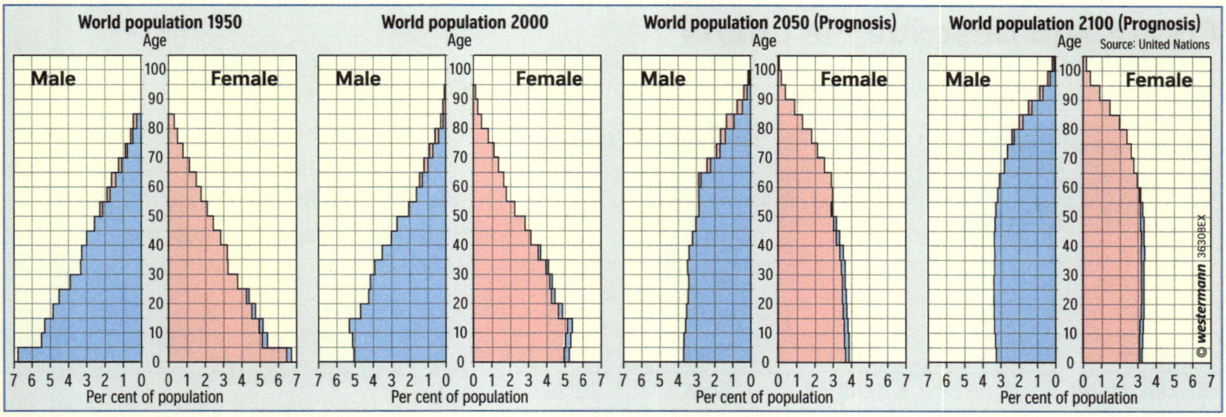

M3 Age-and-gender diagrams of the world population (1950 – 2100)

Model text to evaluate the population structure of the world in an age-and-gender diagram (M1)

Intro-duction

The age-and-gender diagram shows the world population in 2015, **based on data from** the UN.

Evaluation general

The diagram looks like a pyramid, turning into a bell. This means that, on average, **the world birth rate is now medium, mortality is medium to low** and **life expectancy is increasing**. In 2015, we had a growing world population with **a low to medium dependency** ratio.

Additional observations

There is either **a bulge in the age group** between 20 and 29 years or – more likely – an indent in the age groups below. **It is possibly due to** the implementation of the one-child policy in China. So **the number of children** worldwide **may have decreased** after that, as China has a huge population.

Outlook

It can be assumed that the world population will continue to grow in the near future, as **life expectancy will increase**. However, **birth rates seem to be decreasing** worldwide, so that **population growth will** almost certainly **slow down**.

Death rate (mortality rate): number of deaths per year, given in per 1000 of the population. High death rates are above 12 per mille, low death rates are below 6 per mille. Average world death rate (2015): 7.74. A high mortality rate leads to big decreases from cohort to cohort.

Fertility rate: number of live births per female. For a stable population there should be 2.1 births per woman on average. World average (2015): 2.45. A high percentage of women in the reproductive age lead to a higher fertility.

Life expectancy: the average number of years that a person of a population who is born today will live. A life expectancy below 55 years is low, whereas a life expectancy above 75 is considered high. Females have a higher life expectancy than males. World average 2015: 71.7 years (69.6 males, 73.8 years females). The higher the life expectancy, the more the age-and-gender diagram looks like a bell.

Dependency ratio: the percentage of (= the dependent) people below the age of 15 (= children) and above the age of 65 (= aged; elderly) in relation to people between 15 and 65 (= labour force) who are supposed to earn the livelihood for the dependent population. A dependency ratio below 12 is low, above 22 is high.

Population growth: is the result of the difference between births and deaths plus the difference between in-migration and out-migration, given in per cent of the total population per year. Worldwide it is 1.2 per cent / year (2015). There are 24 countries worldwide with negative population growth. These countries show a narrow base in the age-and-gender diagram, so that the overall shape is similar to an urn or onion.

IMPORTANT DEFINITIONS

Birth rate: number of live births per year given in per 1000 (per mil) of the population. High birth rates are above 20 per mille, low birth rates are below 10 per mille. Average world birth rate (2015): 19.73. A high birth rate leads to a wide base in the diagram. Normally, more boys are born than girls.

100800-082, -246 schueler.diercke.de
DE-190 www.diercke.com

Create and Describe Pie Charts

Population of New York City (2016)	Absolute figures	Per cent	Degree
Source: census.gov (2017)	(Million)	[%]	[°]
Bronx	1.46	17.10	61.55
Manhattan	1.64	19.20	69.13
Queens	2.33	27.28	98.22
Brooklyn	2.63	30.80	110.87
Staten Island	0.48	5.62	20.23
Total	8.54	100.00	360.00

M1 Population of the boroughs of New York City

Pie charts or circle graphs get their names from the circular shape – like a pie. They show data in an easy-to-understand way. Pies are divided into slices, pie charts into sectors.

These sectors always add up to 100 % (per cent) and 360° (degrees). The pie chart in **M2** has been created using the data from the table **M1**.

SKILLS HACK

How to convert absolute figures to degrees in a pie chart

General	Part x 100 / Total = Per cent	Per cent x 3.6 = Degrees
Example Bronx	1.46 * 100 / 8.54 = 17.10 %	17.1 * 3.6 = 61.56°
In words	'The population of the Bronx times one hundred divided by the total population of New York City equals seventeen point one per cent for the Bronx.'	'The percentage for the Bronx multiplies by three point six gives the sector for the Bronx with an angle of sixty-two degrees.'

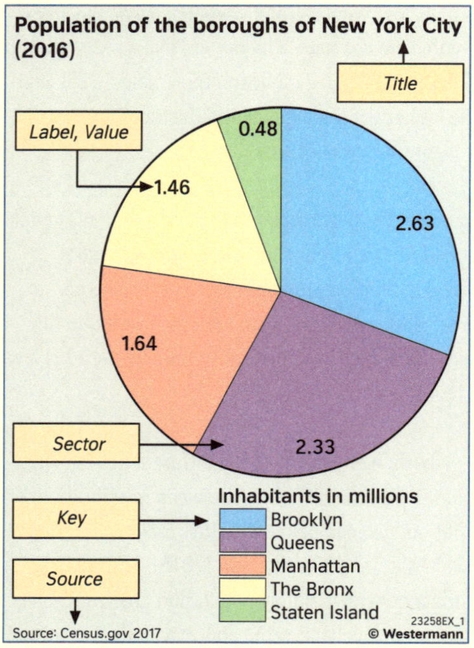

M2 Elements of a pie chart

How to draw a pie chart

STEPS

1. Plan your pie chart: You should not create too many sectors. If sectors are too small to draw, group them into one sector called 'Others.' Always leave this sector to last.
2. Draw a circle.
3. Using a protractor, start plotting from the top clockwise. In general, begin with the largest sector.
4. Add the degrees for the second largest sector and mark it off. Then plot the other sectors from large to small the same way. Finish off with the 'Others' sector (in case you have one).
5. Colour or shade the sectors.
6. Add a title, name the source and make a key (or label the sectors).

Sometimes you find pie charts of different sizes next to each other, called sized pie charts. In these cases the diameters represent the different absolute figures on which the charts are based.

GEO SKILLS

The following model text combines information from the atlas with the description of the pie chart.

Model text to describe the population of the boroughs of NYC from the pie chart (M1)

Introduction

The pie chart shows the population distribution in New York City in the year 2016. The city consists of five boroughs: Brooklyn, Queens, Manhattan, the Bronx, and Staten Island.

Description

The two most populous boroughs are on Long Island. Brooklyn **has got the largest** population of the five boroughs. **Nearly every third** New Yorker lives there. **More than a quarter** of the New York City population lives in neighbouring Queens. Manhattan on the other side of the East River accommodates **just under one fifth** of the New Yorkers, whereas the Bronx is home to about **one in six** of the city's inhabitants. Staten Island, on the west side of the Hudson River, is the least populous borough in New York City. **Less than** six **per cent** of all New Yorker call this borough their home.

Summary

Nearly sixty per cent of New York's population lives on Long Island and **a little over one third** on the Manhattan peninsula.

SKILLS HACK

Dos and Don'ts when you talk about percentages

Working with percentages is different from working with absolute numbers (› p. 17, › p. 20). You should not use certain words which mean absolute numbers when you talk about percentages.

Here are some 'Dos and Don'ts' to observe when you describe per cents:

Do not say...	Rather say
• number • amount	• percentage • part • share
• more - less / fewer	• a higher / lower percentage

100800-219 schueler.diercke.de

DE-025, -143 www.diercke.com

HELPFUL WORDS AND PHRASES

to describe pie charts:

(Only) A	• minority • majority • small / high percentage	of ...
The	• highest • lowest	• percentage of ...
The	• biggest / largest • smallest	• part of ... • share of ...

for more detailed descriptions:

• Exactly	(10 % or 36°) – one in ten ...
• Precisely	(12.5 % or 45°) – one eighth of ...
• Just over	(14 % or 51°) – every seventh ...
• Just under	(17 % or 60°) – every sixth
• Nearly	(20 % or 72°) – one in five
• Less than	(25 % or 90°) – a quarter of
• More than	(33 % or 120°) – a third of
• As much as	(50 % or 180°) – half of ...
• About	(67 % or 240°) – two thirds of ...
• Around	(70 % or 252°) – seven out of ten
• Roughly	(75 % or 270°) – three quarters of ...
• Approximately	

for comparisons:

- (not / not quite) as large / much as ...
- smaller than ...
- larger than ...

Create and Describe Flow Charts

One reason for desertification

Desertification results from a combination of various factors, for example, unreliable rainfall and population growth.

To ensure a constant water supply, people started drilling deep wells, often with foreign aid. At the beginningthere was enough water for the herds and the population.

As the number of herds and people increased rapidly, even more water was needed, so more and deeper wells were drilled. The nomads stopped migrating, since the herds could stay near the wells. Too many cattle on too little grassland led to overgrazing and the plants disappeared.

The new deep wells made the water table sink. As a result, the trees could not reach any water and died. As the ground was no longer protected by trees, grass, and other plants, wind and rain started to carry away the fertile soil. This damaged the soil. The final result was desertification, destroying farmland and pastures – and the process still continues today.

This shows that, in addition to climate conditions, human activities can easily push the ecosystem out of balance in this fragile zone. It is difficult to reverse this process.

M1 Text marked for a flow chart

A flow chart illustrates a process in graphic form by showing the starting point and the different aspects of this process, as well as the results (**M2**).

Flow charts are used in many areas to show processes and the forces involved. Geographers often visualise complicated problems in flow charts.

How to turn texts into flow charts

STEPS

1. Read the text carefully and find out which process you will concentrate on.
2. Find a title for your chart.
3. Mark the most important steps concerning the process.
4. Detect the starting point and the end result of the process. There might be cycles where the result directly affects a starting point.
5. Underline the phrases which show the connection between the key points.
6. For a sketch, write the key points in boxes. Leave space between them. Alternatively you can cut out the boxes and move them around.
7. Connect these boxes with arrows or lines. Make clear what the arrows between the boxes mean. Label the arrows if necessary. Use the terminology from the text.
8. Arrange everything in a clear order and draw your final version.

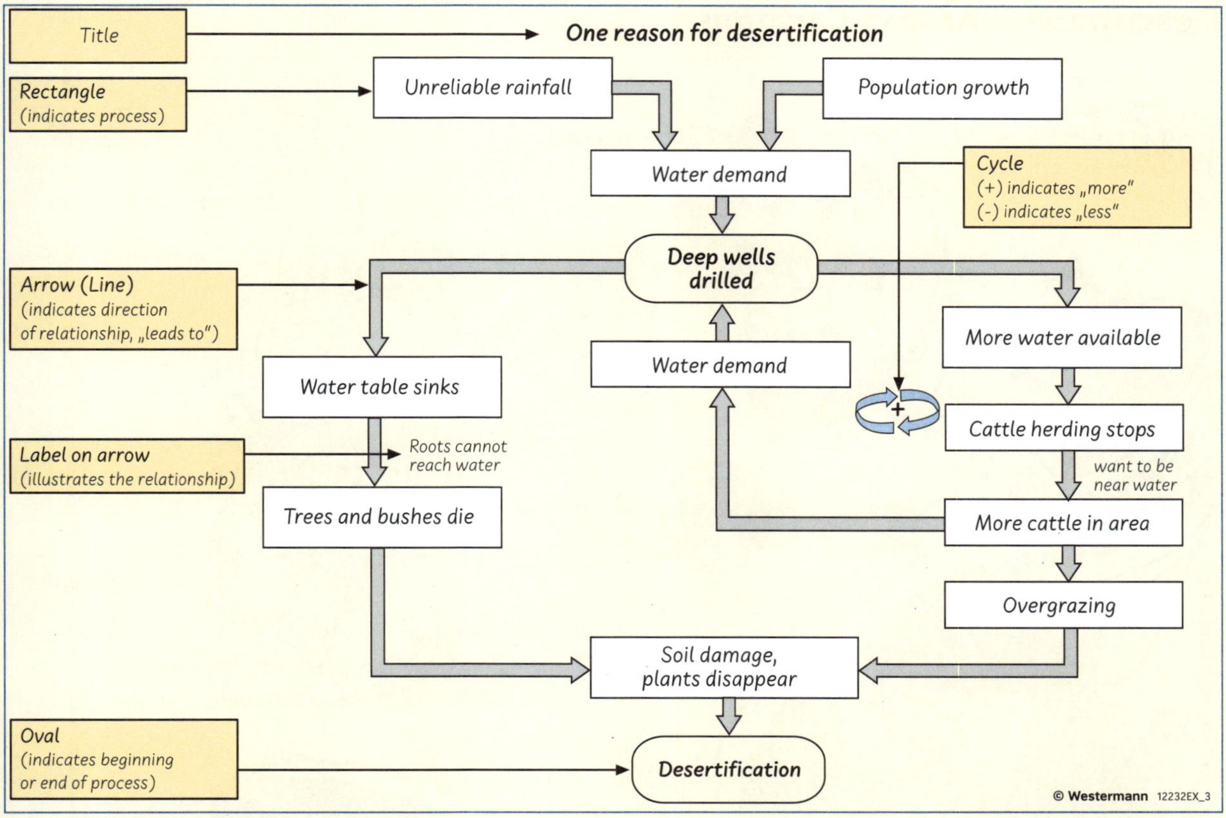

M2 Elements of a flow chart

Model text to describe a flow chart about desertification (M2)

Intro-duction

The flow chart shows one aspect of deserti-fication. **It illustrates** what may happen when people build new deep wells in the Sahel.

Description

When a new deep well is built, **more** water is available for people and animals. So the nomads can keep more cattle and this way even more animals are brought to the wells. More water is needed, leading to more wells being built. **In any case**, too **many** cattle **cause** overgrazing. **As a consequence** the soil is damaged and the plants **disappear**. **Another effect** of the deep wells is that the water table sinks. **This results in** the death of trees and bushes. **Due to this** even more plants disappear and the soil is **further** damaged.

Conclu-sion

The final result is desertification. Valuable land is lost.

HELPFUL WORDS AND PHRASES

to describe flow charts:

- The flow chart shows / illustrates / analyses (the step/steps) ...
- The starting point is ...
- At the beginning ...
- This cycle magnifies / reduces the process so that ...
- It shows the process of ... / different aspects of ...
- This affects ...
- It helps to understand ...
- Finally / In the end ...

linking cause to effect:

- due to ...
- as a result (of) ...
- a result is ... / it results in / the final result is ...
- the end product / by-product is ...
- the outcome is ...
- as a consequence ... / this leads to ...
- because of (this) ...
- when ... then ...
- therefore ... / so ...
- ... is caused by ...

Describe and Assess Cartoons

Date
Cartoonist
Source

Label

Symbol

Title (Caption)

M1 Elements in a cartoon

Kicking the world around

How to analyse a cartoon

STEPS

1 – Content and source
- What are the title (caption), author, publisher, location and date?

2 – Description ("What?")
- What is the event/issue that inspired the cartoon?
- What is the setting, i.e. where and when does the action take place?
- Are there any real – though caricatured – people in the cartoon?
- What is their mood, i.e. happy, sad, angry, etc.?
- Are there symbols/labels/speech bubbles in the cartoon?
- What does the cartoonist exaggerate, i.e. physical features, clothing, positioning and size of figures, facial expressions, and body language?

3 – Interpretation ("Why?")
- What are the connections of the cartoon to the time and place of publication?
- What are the connections of the drawing to its title/the text of the labels/speech bubbles?
- What do the symbols stand for?
- What is the message of the cartoon?
- What is the cartoonist's intention?

4 – Stating your opinion
- What is your opinion about the topic portrayed in the cartoon?
- Do you agree/disagree with the message of the cartoon?
- Are there any open questions you like to raise?

Cartoons are often used to present a strong visual message or point of view on a topic of current interest. As they are visual, rather than textual, cartoons offer lots of information at a glance.

Cartoons use a number of methods to convey their message. By exaggerating features and weaknesses they show people and things in a humorous way. Caricatures (exaggerated drawings in which some features are distorted) and symbols are often included. A symbol stands for some other thing, person, or idea, like the globe in **M1** stands for our planet.
To interpret a cartoon systematically, it is useful to answer the questions on page 30 step by step, taking notes for an outline.

Model text to describe and analyse the cartoon "Kicking the world around" (M1)

Intro-duction

The cartoon "Kicking the world around" **was created by** Jeff Parker **and published in** "Florida Today" in 2004.

Description

It shows a soccer scene with two teams of three players each kicking at a ball which looks like our Earth. The players are fighting and look happy and aggressive at the same time. **This is underscored by** their stupid facial expressions and their oversized, strong bodies. The soles of their shoes have pointed spikes. The players' shirts show the words "Overpopulation", "Pollution", and "Loss of Rain Forests" in one team, "Global Warming", "Fossil Fuel Dependency", and "Species Extinction" in the other one.

Interpretation

The soccer players **represent** major environmental problems of our time. **By oversizing the** players and making them violently kick against a small ball which symbolises a vulnerable Earth, **the cartoonist creates a scenario that shows** our Earth being constantly attacked and threatened ("kicked around") by these problems which are created by aggressive people who enjoy what they do.

Opinion

From my point of view, the cartoon impressively depicts the dangerous situation of our Earth. Our Earth is at the mercy of these problems. **I think it is a good idea to** personalise these problems, because they are created by people.
I share the view presented in the cartoon that we do it without much thinking and aggressively follow our own interest without regard what might happen with our planet.

HELPFUL WORDS AND PHRASES
to describe and analyse a cartoon

Source etc.:
- The cartoon has the title "..."
- The cartoon was created by ...
- The cartoon was published in ...

Description:
- The cartoon shows / is about ...
- In the cartoon you can see ...
- The scene depicts ...
- In the foreground / centre / background ...
- It is a caricature of ...
- The shape / size is distorted / exaggerated ...
- The gesture implies ...
- The body language shows ...
- The clothes appear to be ...
- The facial expression indicates ...
- The speech bubbles / labels say ...
- to draw attention on / to deal with / to illustrate / to represent / to symbolise ...

Interpretation:
- The message of the cartoon is / the cartoon conveys the message ...
- This is a metaphor for ...
- The cartoonist emphasises the idea ...
- The cartoonist wants to express ...
- The cartoonist's intention / message is to ...
- to suggest / to criticise / to ridicule / to expose / to make fun of / to contribute to / to stand for ...

Opinion:
- I agree in principle, but ...
- I (cannot) share the point of view ...
- As far as I am able to judge ...
- In my opinion / from my point of view ...
- (in)appropriate / (un)convincing / (un)biased / (un)realistic / exaggerated
- to support / to (dis)agree / to (dis)like / to dismiss / to reject

Learn with your Textbooks

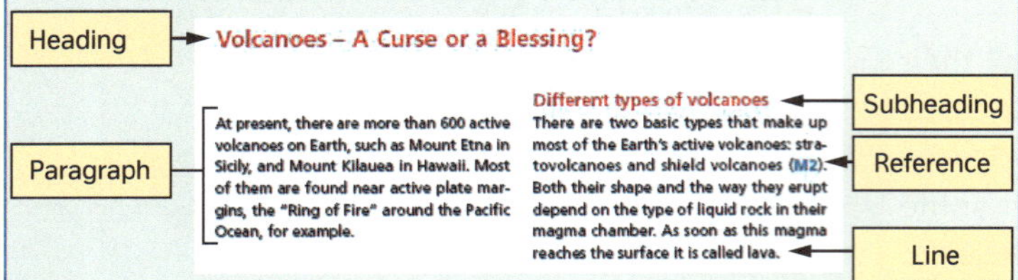

Small letters	a b c d e f
Capital letters	A B C D E
Figures	1 2 3 4 5
Numbers	12 3 456
Punctuation marks	, . : ! ; ?

23297EX
© Westermann

M1 Some elements of texts

Iceland – Plates moving apart

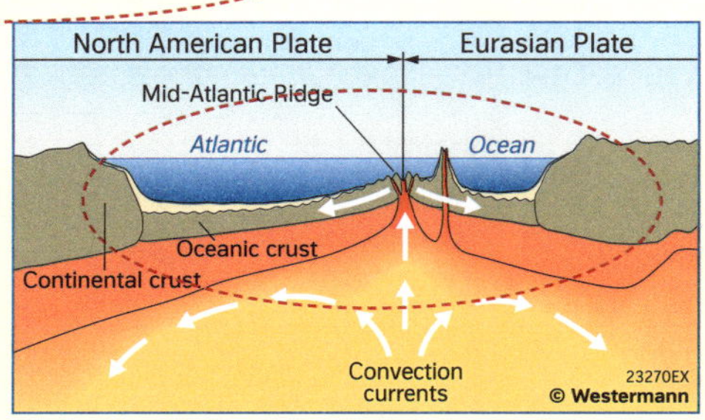

When two plates move away from each other, the result is a gap that is continuously filled by lava coming from the inside of the Earth. As soon as the lava reaches the surface, it starts to cool and turns into solid rock. This process can be found, for example, in the middle of the Atlantic Ocean.
Here, the lava has been building a vast submarine mountain range, the Mid-Atlantic Ridge. In places, this ridge can be high enough to reach above the water surface. This is how Iceland, a volcanic island, was created. Such plate margins are constructive: new oceanic crust is created and the ocean grows. This process is called seafloor spreading.

M2 S – Survey of a textbook page

1. Why is Iceland important?
2. When and why do plates move?
3. How is magma like lava?
4. What are convection currents?
5. Where are the mountains in the middle of the Atlantic Ocean?
6. How can plates move?
7. What is seafloor spreading?

M3 Q – Questions about the text

The SQ3R active reading technique

Every day in school you are asked to read texts. Getting information from texts is therefore as important as it is difficult.
One very successful way to learn from texts is active reading. This means that you not only read the text, you do something with it, copying, marking, asking questions, writing.
The best-known active-reading technique for texts is the SQ3R method. It makes understanding texts more effective. It will take time at first, but you will become much faster over time!

1. S – Survey

Get an overview and a first impression of the material. In our case, it is a text and a diagram (**M2**).
Read the title, look at the visuals and the tasks. Skim the text. Connect your background knowledge with what you read.

2. Q – Question

Ask questions about the material. This is probably the most important step. Make sure that you know what is important to learn from the material given. Ask: What? Who? Why? When? How? Ask questions about the key terms. Write down your questions (**M3**).

3. R – Read

Active reading takes time. Copy (in your own book you can mark) and look up the unknown words and key terms in the glossary, the register, or other sources. Write down (or highlight) key terms and their connections (**M4**). Find answers to your questions in the text. Take notes. If you cannot answer a question yourself, discuss your problems with others. When new questions arise, add them to the others. Never be afraid to ask. After 'Reading' you should have found answers to all your questions.

4. R – Recite

Close your material, recall your questions and answer them in your own words. This way you can show yourself that you really understand what you write or talk about. Reciting can be done in ways other than writing a text in the form of a summary. You can do a list, a mind map (**M5**), or a drawing to answer your question.
If you cannot answer your question(s), read through the material again.

5. R – Review

You cannot expect to answer each question correctly after the first reading. Reviewing means that you go through your notes after some time. You can highlight key information.
Organise the answers to your questions in a different way: make a list instead of a summary, turn a mind map into a text, make index cards, and so on.
You can review with a partner, where the partner reads out the question and you answer. This way you can check if your answer is correct and that you phrase it in a logical way. Your partner might have different questions which you can try to answer.
Review regularly, not just before a test.

HELPFUL WORDS AND PHRASES
to describe facts from texts:

- The text is about / ... deals with ...
- The text tells about / ... gives an account of ...
- It says / states (that) ...
- It claims / suggests (that) ...
- We can sum up the (first) paragraph by saying that ...
- As a consequence / result ...
- Thus / hence / therefore ...
- (only orally:) I cite: quote ... unquote.

Iceland – Plates moving apart

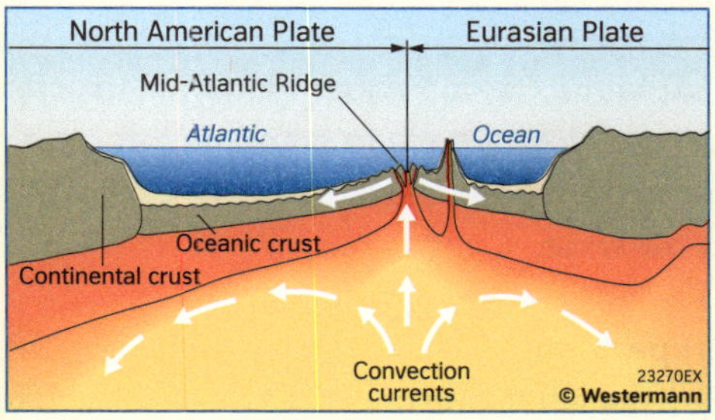

When two plates move away from each other, the result is a gap that is continuously filled by lava coming from the inside of the Earth. As soon as the lava reaches the surface, it starts to cool and turns into solid rock. This process can be found, for example, in the middle of the Atlantic Ocean.
Here, the lava has been building a vast submarine mountain range, the Mid-Atlantic Ridge. In places, this ridge can be high enough to reach above the water surface. This is how Iceland, a volcanic island, was created. Such plate margins are constructive: new oceanic crust is created and the ocean grows. This process is called seafloor spreading.

On constructive plate margins, lava from inside of the Earth (Earth's mantle, see page 8) fills gaps between plates, turns into solid rock, and creates mountain ranges.

M4 R – Read (The text and its summary after the SQ3R method)

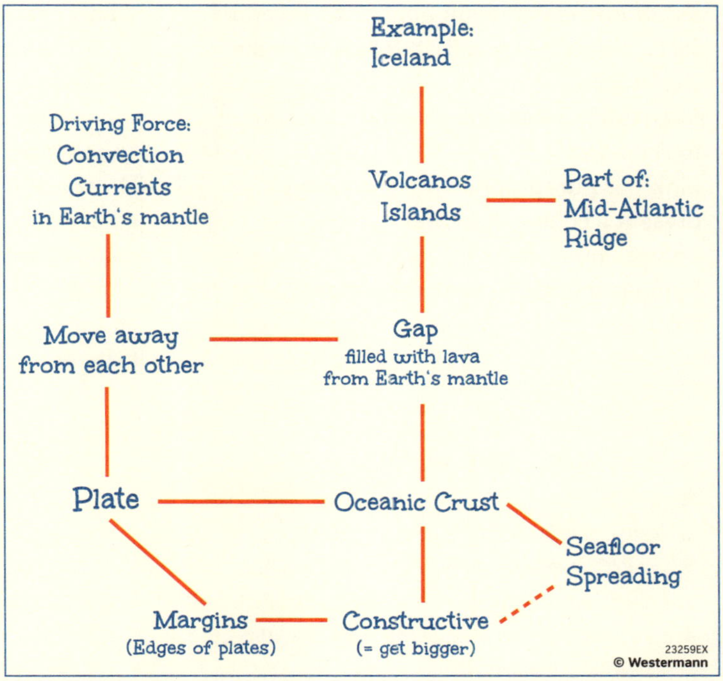

M5 R – Recite (Here with the help of a mind map)

The Instructions "Explain", "Comment on" and "State your opinion"

Structure of a solution with the instruction 'Explain'

„Explain ..." is a frequent exam instruction. Generally, you have to combine information from the given material with background knowledge.

Preparation
Find out what kind of explanation is asked for: Do you have to explain causes, reasons, conditions, effects, results, consequences, or purposes? Determining this is an important step to help you not to go astray with your answer.

Introduction
Repeat the topic and task in your own words

Main part
Each argument should start with a factual statement or short description, followed by the actual explanation.
Depending on the task, explain by giving a reason, by naming the causes, by pointing out the effects, etc. Link fact and explanation with a word or phrase from the wordlist. If one explanation applies to several facts, name the facts first, followed by the explanation.
Suggested paragraph structure: Fact(s) – explanation.
Link the paragraphs with **transitions** to present a coherent answer.

Conclusion
Sum up the main points and give a final comment.

Task: Explain the likely problems of oil transport, using your atlas.

Model text

Crude oil is transported from its production site to its consumers, mainly through pipelines and by huge oil tankers. The transport is not without dangers.

First and foremost, each leak that happens during transfer can harm the environment because oil is poisonous and flammable.

Moreover, when oil comes into contact with water it can spread easily. Thus, it can contaminate huge areas quickly and even affect the ground water.

In addition, production sites often are in politically unstable regions, so pipelines and storage tanks can become targets in warfare and terrorist attacks.

Furthermore, the atlas (264,1) shows that many production sites are in remote regions of the world so that environmental damage remains unnoticed for quite some time.

Above all, transport across the oceans can be dangerous. As a look in the atlas (252,2) confirms, most tankers have to pass through areas with frequent storms. When tankers are damaged or sink, then oil is released which kills the sea life around it and sticky oil residues damage coastal ecosystems.
Tourist regions suffer **too**, because beaches are affected by tar from oil spills.

Last but not least, oil is flammable, so there is a risk that along the transport routes burning oil will endanger human lives.

All this shows that the transport of flammable and poisonous oil endangers ecosystems and humans alike. Therefore, oil should be transported with great care.

HELPFUL WORDS AND PHRASES
to explain

... a cause or reason: as; because (of); being that; due to; for; hence; in that; in view of (the fact); owing to; since.
... a condition: as long as; even if; given that; granted (that); if; in case; in the event that; on (the) condition (that); only if; provided that; providing that; unless.
... a consequence: as a consequence; if not; if so; in that case; otherwise; that being the case; then; under those circumstances.
... an effect or a result: accordingly; as a result/an effect (of this); because of this; for this reason; hence; so; so much (so) that; so that; therefore; thus.
... a purpose: for; in order that; in order to; so (that); to the end that; with this in mind.

Structure of a solution with the instructions 'Comment on' and 'State your opinion'

Preparation
Think about the topic. What is your opinion? Jot down notes about arguments to support your ideas.
Note: *Write in simple present.*

Introduction
Name the problem you are going to comment on. Give a short overview on what you think.
Main part
Always give reasons for your opinion. Illustrate with examples if possible. You can comment on diverging opinions – usually in a negative way. Concentrate on one main point per paragraph. Link the paragraphs with **transitions**.
Suggested paragraph structure: Opinion – Reason (– Example).
Link the arguments.
Conclusion
Summarise and generalise the most important arguments that best support your opinion.

Task: State your opinion on fracking.

Model text

Over the last few years, fracking has become an important means of recovering oil and gas. In my opinion it should be stopped immediately.

First of all, I fear for the environment, because there can be irreversible damages to vital water reserves. We have seen groundwater polluted by oil and gas as a result of fracking.

I have read that the supply of oil and gas leads to dropping prices. This would reduce our heating bills and transport costs. As far as I can see, that is only part of the truth, as cheap petrol and gas leads to higher consumption, which again leads to an increase of harmful pollutants and greenhouse gases. Moreover, finite resources will be depleted more quickly this way.

Finally, fracking causes earthquakes. Until now the tremors had been small, but to me this proves that fracking might trigger major quakes with catastrophic results.

There is no doubt that the environmental impact of fracking is tremendous. In my opinion, governments should advise the drilling companies to stop fracking at once.

HELPFUL WORDS AND PHRASES

... to state your opinion:
as far as I know / understand / can see, ...; from my point of view ...; I am (not) sure / certain / convinced that ...; I am of mixed opinions about / on ...; I have read that ...; I think / believe / suppose ...; in my experience ...; in my opinion, ...; it seems to me that ...; my view is that ...; to me, this proves that ... .

... for linking arguments:
additionally; again; along with; also; and,; another; another aspect/point is that ...; as well; besides; equally important; finally; first(ly); first of all; further; furthermore; however; in addition; in contrast; lastly; likewise; moreover; nevertheless; next; otherwise; second(ly) etc.; similarly; too.

... for giving examples:
for example ...; for instance ...; in the case of ...; in this case; look at ...; to demonstrate; to illustrate; when / where ... .

... for the conclusion:
as (I) already indicated ...; as a result; as shown; consequently; as a consequence; finally; hence; in conclusion; in summary; on the whole; so all in all ...; summing up; therefore; thus; to conclude; to summarise; (weighing the pros and cons,) I come to the conclusion that ... .

SKILLS HACK

Educated guess
An educated guess is an assumption you give based on observation in the material and your background knowledge.

HELPFUL WORDS AND PHRASES
... to use in an educated guess:
to guess; to think; to assume; to suppose; to expect; to believe; (quite, very) likely; (almost) certainly; possibly; not unlikely; presumably; In my opinion; If ... then...; Based on ...; I predict that ...; I guess that ...; The material suggests that ... .

The Instructions "Compare" and "Contrast"

Structure of a solution with the instruction "Compare" or "Contrast"

In "Compare ..." you should state similarities and differences, whereas in "Contrast ..." you should focus on the differences. There is no interpretation of the findings required in these instructions unless specifically asked for.

Preparation
Make a list like **M1** where you jot down notes about similarities and differences of the two things you compare.

Introduction
Chooses what will be compared / contrasted and identifies the major lines of the comparison.
Main part
Presents each of the compared points in a separate paragraph.
- Short description item 1
- Short description item 2
- Comparison / Contrast

(in "Contrast" it is possible to combine the "Short description item 2" with naming the differences)
Summary
Summarizes the key similarities and differences of the two compared materials.

Task: Compare the population development of Buffalo and Boston between 1950 and 2013 (**M2**).

Model text
I want to compare the population development of the two Rustbelt cities, Buffalo/NY and Boston/MA between 1950 and 2013 according to data from the US census website.
In 1950 the population of Buffalo was 580 thousand. Since then the population declined permanently until dropping to 259 thousand inhabitants in 2013.
Boston had a population of 801 thousand in 1950. The population size shrunk to 563 thousand in 1980. From then on the number of people in Boston increased more and more until the city reached 646 thousand inhabitants in 2013.
Both cities lost many inhabitants from 1950 until 1980, Boston nearly every fifth, Buffalo even every third inhabitant.
But from 1980 onwards the population development differed. As Buffalo lost more than a quarter of its remaining population over the following 33 years, Boston reversed its trend. Unlike Buffalo, Boston's population started growing, even with increasing speed, to its population size of 2013.
So according to the latest figures, Boston's population is again rising, whereas the number of inhabitants in Buffalo continues to sink.

Things to be compared/ contrasted → Facts/Details ↓	Population in Buffalo	Population in Boston
Similarities	– 2013 Less Population than in 1950 – Both decrease until 1980	
Differences ↓	– Loss of >50%	– Loss of 20% ↓

M1 Graphic organiser for "Compare" / "Contrast"

Year	Inhabitants (in 1000)	
	Buffalo	Boston
1950	580	801
1960	533	697
1970	463	641
1980	358	563
1990	328	574
2000	293	589
2010	261	618
2013	259	646

Source: www.census.gov

M2 Data for comparison

HELPFUL WORDS AND PHRASES
for differences:
a (clear, distinct, striking) difference; although; another distinction; but; contrary to; despite (this fact); to differ (from); different from; distinction; distinct; distinctive; divergent; diverse; even though; further; however; in (sharp) contrast (to this); in spite of; instead; nevertheless; nonetheless; on one hand; on the contrary; on the other hand; opposite; otherwise; rather; still; the reverse of; the next / third / ... distinction; though; to the contrary; unlike.

Structure of a solution with the instruction "Discuss"

In the instruction "Discuss ..." you are asked to balance arguments of statements or positions against each other. A clear and logical sequence which reflects your stand is essential. The instruction "Discuss" is targeted. It is supposed to end in a conclusion with your decision or verdict on the matter.

Introduction
(Hook to engage the reader)
Background information about the topic (statement, claim, position)
Main part (Arrange arguments (see **M3**) in order of importance)

- Argument 1
- (Link) – Argument 2
- (Link) – Argument 3
- ...

Conclusion
Sums up the main points to support your position, ending in your verdict on the matter.

Task: "Discuss the pros and cons of producing electricity by nuclear energy."

Model text of a main part with three arguments
Electricity by nuclear power is the least expensive one. Economists argue that it is the cheapest way to produce electricity with less than 1 US Cent per kWh. You can hear from environmentalists that this price does not include follow-up costs. You must consider however that no calculation of electricity costs by any source does include everything, so this leads to pure speculation.
Then there is the problem of greenhouse gases. The production of electricity by nuclear power does not emit any greenhouse gases. A well-known claim is that during the production of uranium and other subsequent processes, CO_2 is released. But this is a negligible amount compared to the production of electricity by fossil fuels.
You can well add the aspect of security of supply. Nuclear power produces the required amount of electricity 24/7, whereas the sun and wind energy are very unpredictable and unreliable. The supporters of so-called green energy allege that the surplus electricity can be converted into other forms of energy which then can be turned into electricity once the sun does not shine and the wind does not blow. However, this is very costly and will further add to the price of electricity.
...

Statement, claim, position (Pro or Con) 1
Supportive reasoning 1
Critical reasoning 1
Refutation 1
(Link) – Statement, claim, position (Pro or Con) 2
Supportive reasoning 2
Critical reasoning 2
Refutation 2
...

M3 Two arguments in a discussion

HELPFUL WORDS AND PHRASES
for similarities:
also; another similarity; as; as well as; comparable; comparatively; compare; equal(ly); identical(ly); in (much) the same way; in relation to; like; liken(ed); likeness; likewise; look like; match; matching; relative to; relatively; resemblance; resemble; resembling; same as; similar to; similarity; similarly; too.

HELPFUL WORDS AND PHRASES
"For" arguments:
The main argument for it is ...
What also speaks for it ...
There is also the fact / opinion that ...

"Against" arguments:
Against this it has to be said ...
In / By contrast; on the contrary

"Neutral" links:
Particularly important is ...
Above all; in particular; especially; mostly

Write a Geographical Essay

A geographical essay is a written presentation. So you could prepare for your essay in a similar way (› pp. 40-41).
Your essay should show knowledge and understanding of a particular geographical topic by researching it and writing about it. It often contains an instruction (› pp. 50-51) and one or more key terms.

How to write a geographical essay

STEPS

1. Planning
Develop an essay plan. Structure is important. A good way to do this is to outline the main points as headings. Under each heading write in one sentence the broad point you want to make. After this, go back and consider whether the general argument you want to make follows through the structure. Change the paragraphs in the essay plan until it fits. Now you know exactly which information you have to look for. But you should always be open to new ideas which might come to you during your research.

2. Doing the research
To find the most reliable and appropriate information, it is important to refer to different sources including the Internet, reference books, atlases, school books, text books, journals, newspapers, and magazines. There is no substitution for reading widely. Internet sources not always contain reliable information, so they need to be checked.
When making notes, the direct copying of another writer's material, should be in single quotation marks. If you use it in the essay, you should paraphrase or reorganise it. All collected information should be referenced within the report and in a bibliography (5). Once all information has been collected, it needs to be reviewed and sorted into the order you already outlined. Make corrections in the order if necessary.

3. First draft
Once all of the information has been collected, it needs to be reviewed and sorted into the logical order you already outlined in the planning stage.
Why not develop and structure your ideas using the P.E.E.L. method (**M2**)?
The general structure of the essay would then be: Introduction – Main part (P.E.E.L. 1, P.E.E.L. 2, ... P.E.E.L. X) – Conclusion.

General rules for writing the essay

Introduction
- Set the topic in a wider context.
- Or provide definitions of the key words and your interpretation of the task.
- Or give an outline of the essay.

Read your introduction again after you finished the essay to see if your way of introducing the topic still fits.

Main part
- Structure your argumentation, following the headings with the key ideas from your planning.
- Constantly refer to the essay topic to make sure that you do not stray.
- Put each key idea in one paragraph.
- As a rule of thumb, assume that the reader has little knowledge of the topic.
- Use maps, graphs, charts, diagrams, and pictures carefully as a mean of supporting your argument. You need to refer to the media in your text and explain the meaning.
- Avoid stereotypes.
- Quotations should only be used to support or expand an idea that you made in an argument.
- Accurately reference all the ideas and materials that you have used from other sources. A failure to do so is called plagiarism. It is a serious offence.

Conclusion
- Draw out the main points of the essay.
- Or give a clear sum-up answer or verdict to the essay task.
- Or comment on the bigger picture and provide some predictions for the future.

Planning →	P.E.E.L. structure	→ Draft of the paragraph in the essay
Heading: Greenhouse gas CO_2. Broad point: Show the relationship between CO_2 and temperature		*(Previous P.E.E.L. paragraph)*
	Point	CO_2 contributes to global warming.
	Evidence	Over the past 150 years the average temperature of the Earth has increased as has the content of CO_2 in the atmosphere.
	Explanation	CO_2 lets the visible light pass but reflects infrared radiation – heat – which is emitted from the Earth's surface. Thus the molecules function like the glass panes of a greenhouse. Therefore CO_2 is called a greenhouse gas.
	Link (to main topic and to next paragraph)	However, there are other gases which contribute to global warming in a similar way.
		(Next P.E.E.L. paragraph)

M1 Developing a paragraph in an essay with P.E.E.L.

STEPS (CONTINUED)

4. Editing

Produce second or even third drafts. It is a good idea to read your essay aloud to uncover small errors. Moreover, it helps to improve the flow of the essay. Do not totally rely on an automatic spell checker as many words spell similarly but mean something different (f. e. their – there). Spelling should follow the agreed language – British or American English.

5. References

At the end of your essay you have to provide a bibliography which includes all the material cited within the text and that material which had been consulted in the preparation of the essay. It has to be ordered in alphabetical order of the surname of the author. You find some important hints for referencing for a bibliography on this page.

In-text reference should be "(surname, year of publication)" if you only use the idea of an author. If there had been more authors, name them all. A direct in-text reference should be in 'single quotation marks', followed by "(surname, year of publication, page #)".

SKILLS HACK

Quick guide to correct citations

Book:

Generalised: Name of author(s) (a. o. = and others): Title. Publisher, Year, Page(s) from which information was retrieved.

Example: Sarah Flöck a. o.: Diercke Geography, Volume 2. Westermann, 2017, pp. 74-75.

Newspaper article:

Generalised: Name of the Author(s): „Title", Name of newspaper, Date (YYYY-MM-DD), Page(s).

Example: Aurelien Breeden: "Waste Water is Heating Pools in Paris", New York Times, 2016-12-13, p. A6.

Magazine article:

Generalised: [unknown author(s), therefore none given] "Title", Name of magazine, Number and Year, Page(s).

Example: "Hull of a wind behind it", The Economist, 16–23 Sep 2017, p. 46.

Website:

Generalised: Name of the Author(s): "Title" of the website (in this case the main heading), Web address, Date retrieved (YYYY-MM-DD).

Example: Jonn Elledge: "London's Tube has been running so long it's literally raising the temperature of the earth around it", http://www.citymetric.com/transport/londons-tube-has-been-running-so-long-its-literally-raising-temperature-earth-around-it, 2017-09-22.

Prepare a Talk or a Presentation

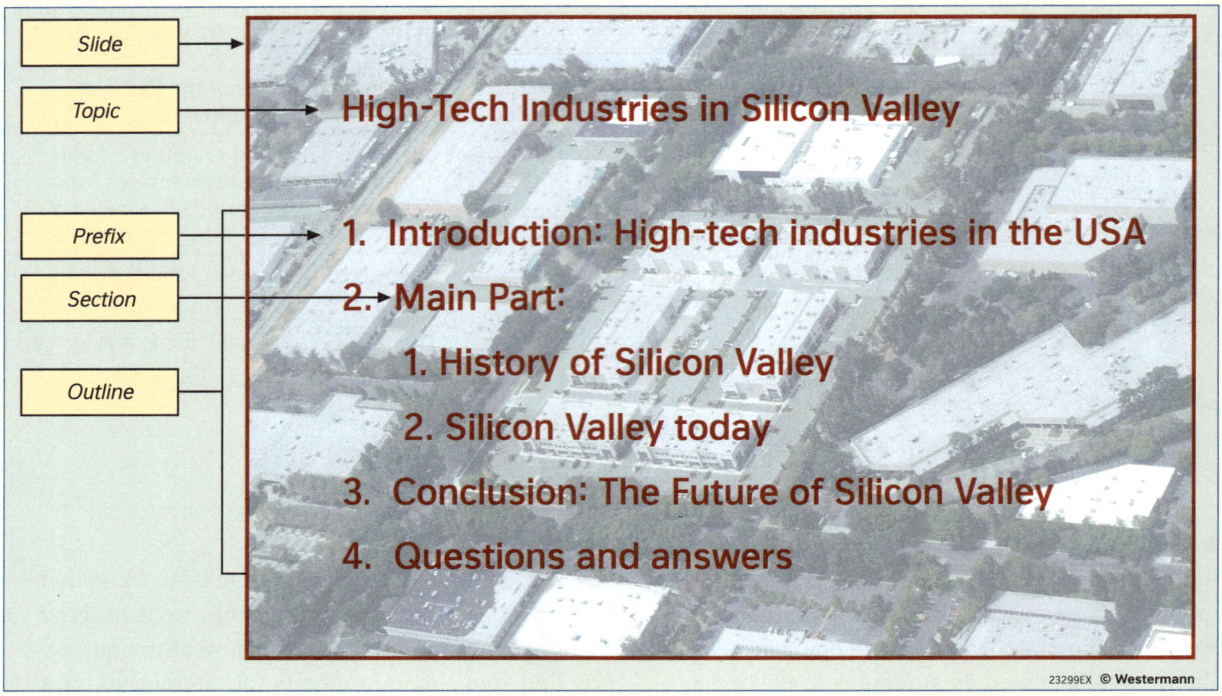

M1 Elements of the outline of a presentation on the topic of 'High-tech industries in Silicon Valley'

Preparing for a talk or presentation is similar to preparing to write a geographical essay. So many of the tips given there (› pp. 38-39) will also help you in the preparation of your talk.

How to prepare for a talk or a presentation:

STEPS

1. Make yourself an expert on your topic:
- Collect information from several sources. Keep a list of the sources you used.
- Double-check the information, especially if you have it from the Internet.
- If you work in a group: Decide on who does what in your group. Share the work evenly.

2. Organise the material:
- What is important?
- What are the main points?
- Which structure will my talk have (see **M1**)?

3. Prepare the talk:
- Make it interesting and motivating.
- Write down key terms.
- Use your own words, never copy text from the material.
- Use unknown vocabulary sparingly.
- Make a word list with the necessary vocabulary to present in class.
- Don't use too many numbers, use graphs instead.
- Prepare visual aids (computer presentation, sketches, diagrams, transparencies, photos, models ...).
- Check the presentation tools (**M2**) before the talk.
- Check if your software will run on the school's hardware.
- Set up questions and / or a knowledge check to find out what your audience has learned.
- Prepare a handout or written summary if required.
- Practice your talk at home, including the body language.

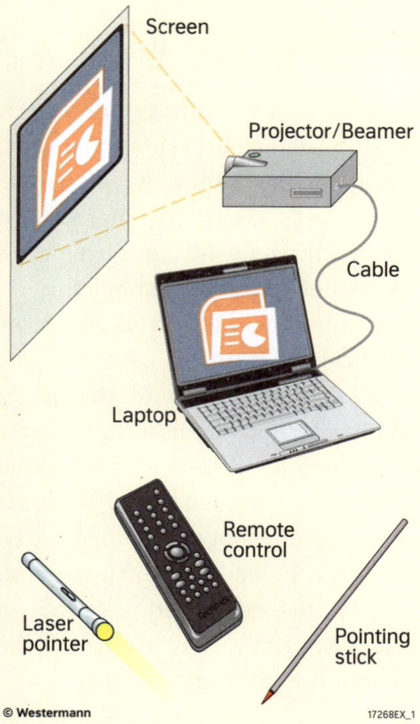

M2 Presentation tools

Give the talk

- Speak slowly and loudly.
- Speak freely, use notes with prompts only.
- Make eye contact with your audience.
- Catch the attention of your classmates.
- Give an outline of your talk (**M1**).
- Give and explain important (key) terms.
- Introduce unknown words.
- Relax and enjoy your presentation.

SKILLS HACK

The quality of your performance depends on the following aspects:

- Opening: Is it motivating, informative, and to the point?
- Description of the topic: Is the topic correctly presented?
- Clarity of details: Has the topic been presented in a logically structured way?
- Background information: How much do you know about the topic?
- Style: What about your command of the language, body language, …?
- Organisation: How is the use of media?
- Duration of the talk: Too long? Too short?

HELPFUL WORDS AND PHRASES

to use in a talk or presentation:

Introduction:

- I want to speak to you about …
- We will give a talk about …
- Our topic is …
- Let me give you a short outline / overview of …

List:

- Firstly … After this … Next … Finally …
- Now I'd like to look at …
- I'll come to …
- I'm going to show you …

Direct attention:

- As you can see …
- You will note that …,
- It is obvious that …

Introduce evidence:

• I'd like you to look at … • Let me show you … • (Let's) (have a) look at …	• this graph. • this map. • this chart. • this diagram. • this photo. • this model.

Explain:

• The rise • The increase • The fall • The decrease	• led to … • was due to … • caused … • brought about (by) … • stemmed from … • resulted in …

Give reasons:

- Owing to …
- Because (of) …
- Since there is …

Interpret information:

- These figures indicate a … / an … / that …
- It becomes clear that …
- Our findings show a … / an … / that …
- This may suggest a … / an … / that …
- I think this all means a … / an … / that …

Move on:

- I now want to go on to …
- This brings me to …
- I shall come back to that later.

Look back:

- Let's go back to …
- I want to go back to …

Conclude:

- To sum up …
- Let me end by pointing out …
- In conclusion …
- I would like to finish by saying …
- And so …

Taking Notes during Talks and When Watching Videos

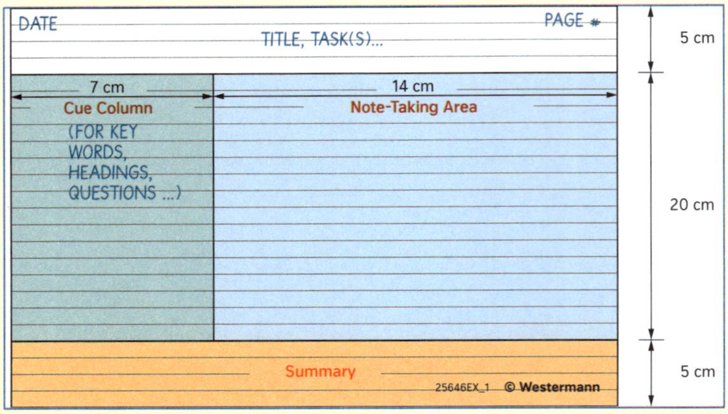

M1 Set-up of a note-taking paper

How to make note-taking easier

STEPS

1. Preparation

An established set-up of a note-taking paper is shown in **M1**. If you are given a task in advance, write it down at the top of the page along with given key words in the cue column. If there are several tasks, use a new page for each task.

2. Note-taking during talk or video

- Use the note-taking column (**M1**).
- Use telegraphic sentences, symbols, and abbreviations (**M2**) to speed up note-taking.
- Bullet points or numbering for a clear structure.
- Leave space for further additions.

3. Note review

After the talk or when the video is over, go through your notes to add things, underline, colour-mark, connect, or cross out unnecessary information. Use the space at the bottom of each page for a summary.

4. Questions

In a different pen colour, formulate questions in the cue column, based on your notes. Writing questions helps to clarify meanings, reveal relationships, establish continuity, and strengthen your memory. Check if you can solve the tasks which were given to you before the talk / video.

5. Revise

Looking at the tasks, questions or key words in the left column only, answer the questions, reestablish the facts or retrace the ideas indicated by the key words in your own words.

A good way to revise is to write several multiple choice questions about the subject for your peers.

Note-taking is a very important skill not only during a talk or presentation given by your teachers or peers but also when watching videos. Like every other skill, you need to practice note-taking, even more so as it is needed for all your subjects.

You take notes because it helps you to concentrate on the subject. Later on, your notes trigger memories of things you already learnt. You can refer to the notes as a resource for review and reference, because often they are easier to understand than your schoolbooks.

Good notes are clear – so you understand them. They are filed properly – so that you can find them easily. They fit you – so they follow your own lines of thought.

Note-taking when watching videos

- Ideally you watch a video in your own time.
- Watch it several times, adding to your notes.
- Use the subtitle function when available. [Caution: In many videos the subtitles are not correct!]
- When watching videos, jot down the time of the video passed for later reference.
- If the video is available online, *http://www.videonot.es/* can assist you taking notes. It is advisable to create an email account especially for this purpose.

Talking, or writing a text, about a video

A video always reflects the view of the people who made or screen it, being fictional or factual. Therefore, if you describe the content of a video you should not state the view expressed in it as facts. Rather phrase it as assumptions or opinions of the person who says it.

When evaluating a video being skeptical is important. So, as a first step, you have to try to distance yourself from what and how it had been said in the video.

+	and, add, plus	ea	each
=	is, are	min	minimum, least
≈	about, approximately	max	maximum, most
>	more than	vs	versus, against
<	less than	qty.	quantity (amount)
#	number	N.B.	note well
x	times	eg	for example
→	leads to, causes	ie	that is
←	comes from, caused by	b4	before
↔	exchange	p	before
↺	change	f	after
/·	compared to	w/	with
/	or, per	w/o	without
↑, ↗	increase, grow, gain	re	regarding
↓, ↘	decrease, loss	b/c	because
♀	femal, woman, girl	s/t	some-thing
♂	male, man, boy	s/o	someone
yrs	years	d/f	different
@	at	govt	govern-ment
!!	important	prbs	problems

M2 Suggestions for a note-taking

USEFUL WORDS AND PHRASES
to describe video contents:
Types of videos:
Animated film (cartoon); broadcast; docudrama; documentary; feature film; film clip; motion picture; (news) report; TV coverage.

People in videos:
Actor; broadcaster; cartoon / central / main / minor / favourite character; host; interviewee; interviewer; lecturer; moderator; presenter; reporter; speaker.
To act; to be interviewed; to broadcast; to demonstrate / show; to host; to lecture; to present; to report; to speak.

Editing:
Off-camera (comment); offscreen; on-camera; onscreen; subtitles; to translate; voice-over.

Opinion:
Allegation; allegedly; assertion; assumption; fake news; presumption; suggestion.
To agree; to argue; to be of the opinion that; to claim; to comment; to disagree; to dissent; to imply; to suggest.

'Describing' phrases:
According to ... (= person in video) ...;
I quote ... (=person in video): ' .... ';
In view of the fact that ...;
The ... (= type of video) is about ...;
The ... (= type of video) deals with ...;
The ... (= type of video) is set in ...;
In one shot/scene, we could see ...;
The scientist /expert remarks that ...;
The filmmaker stresses the fact that ...;
The ... (= type of video) concludes that ... .

'Evaluating' phrases:
It seems to be that ...;
This may be true ... but ...;
This argument / fact is convincing / not convincing because ...;
We need to know more about ...;
The topic is controversial because ...;
I expected that the ... (= type of video) would deal with ...;
The ... (= type of video) leaves the following questions ...;
We have learned that ... .

Prepare for a Discussion or a Role Play

Preparing for a discussion in a group or preparing for a role play is quite similar. When you prepare yourself for a discussion however, you have to take on multiple perspectives, as in a role play, you have a fixed role.
This does not mean that you should not be prepare yourself for other points of view and how to deal with them.

M2 Role play in school

Role play

Role plays in Geography lessons are simulations of real-life situations. They are primarily designed to help you to develop a deeper understanding for certain positions and conflicts, for example in case studies in which environmental and economic or private and public interests clash.

Most role plays follow a similar pattern: An introduction of the topic by the chairperson or moderator will be followed by short introductions of the people taking part in the role play. There they tell who they are, what their positions are, and where their interests lie. Then follows a discussion or interchange of points with or without interaction by the moderator.

At the end the participants give a conclusion and outlook from their point of view and the moderator sums up the results.

How to prepare for a role play

STEPS

- Read the instructions for your role carefully. If necessary, ask your teacher to clarify.
- There is always much information that is not included in the instructions. Read different sources about your role background to make yourself more familiar with the situation.
- If required, prepare your opening and closing statements.
- Prepare a skeleton structure of your performance, but keep everything flexible. Do not formulate a rigid plan and do not expect things to work out exactly as you planned.
- Anticipate the points of the other participants of the role play. Be ready for a discussion (see below).
- If you have to have notes, try to keep them on one sheet of paper. Use bullet-points and write in a large size which is easy to read.
- Practise to become familiar with your role.
- Be aware that your body language is just as important as your words.

M1 Participants of a discussion

Discussion

The centrepiece of many role plays is the discussion where, after having made their statements before, the participants argue with each other.
To be most convincing in a discussion, to have the best arguments, and to persuade the audience of one's position should be the aim of each member of the panel.

Tips for a successful panellist in a discussion

- Be well prepared about the topic. Read about it, covering all angles. This way you have knowledge of opposing opinions and can be ready to react to them.
- Be clear about your position. Write down your arguments and rank them from most to least important to you.
- Distinguish between facts (which can be supported with data), opinion (where other sources can be cited), and claims (for which there is no clear evidence).
- Data can be very persuasive. Be sure they come from reliable sources.
- Use figures sparingly. Too many figures confuse the audience.
- Having a pre-planned opening statement of just a few sentences will help you settle down. If you do not have the chance to give an opening statement, make sure that you can state your major points when it is your first turn.
- Be sure and confident when you state your points.
- During the discussion refer to the statements of other panellists. Object or agree in making your position clear.
- Show empathy or, at least, respect to the view of other panellists and members of the audience.
- Be polite and do not belittle anyone.
- Be able to compromise, in order to find common ground, allies, or partners.
- Make yourself familiar with typical English phrases. You can use them to introduce yourself, state your points, or react to opposing arguments. It is always bad in a discussion to be lost for words.

USEFUL WORDS AND PHRASES IN A DISCUSSION

to state an opinion´:
In my opinion ...; I think (that) ...; I would say (that) ...; It is my view (that) ...; My opinion is (that) ...; It seems to me (that) ...; I feel (that) ...; I (strongly) believe (that) ...; I would say ...; The way I see it ...; What struck me was ...

to suggest something:
Why not ...?; Why don't we ...?; What about ...?; It would be nice to ...; I'd like to suggest that ...; It seems to me we should ...; I have an idea: ...

to refer to something or someone:
With respect to ...; With reference to ...; From the point of view of...; In terms of ...; With regard to ...; As far as ... is concerned ...;

to support and agree:
I (quite) agree.; I think so, too.; Exactly.; It is true.; I think that's a good point.; You're absolutely right.; I couldn't agree more.; Along those (same) lines ...; I would like to add ...

to contradict:
I wouldn't agree with that.; That's not the point here.; I see your point, but I think ...; You must be joking!; It may be true that ... but ...; I'm afraid I don't agree with you there.; I disagree with ...; I don't think so.; On the contrary!; Unfortunately I cannot agree on that point ...; I see the point but ...; I am not sure about ...; I doubt that ...; I suspect that ...; I criticise ...; Well, it depends.

to clarify the position of others:
This isn't that relevant because ...; So are you saying that ...?; So in other words you think ...?; Can you explain what you mean?; What do you mean by ...?; I don't quite understand what you mean.; That's not quite clear to me.

to give in:
OK., I think you're right.; I take that back.; I didn't take ... into account.; You do have a point there.; That all considered, I have to give in.; You've convinced me.

Conduct a Survey

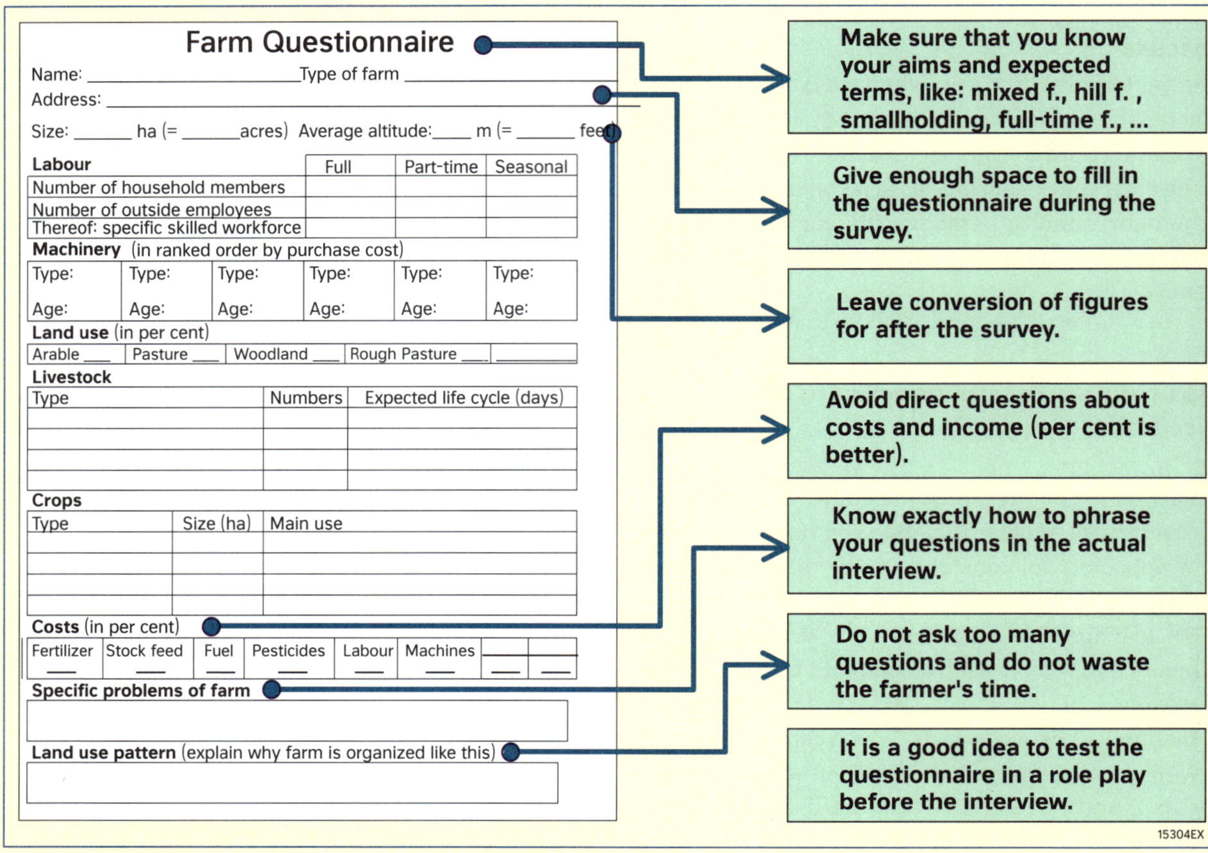

M1 Questionnaire for a farm study

Questionnaires are valuable because they provide up-to-date information not available anywhere else, for example in newspapers, scientific journals or government statistics. They are very often used to find out about peoples' motives and opinions.

To conduct a survey in the English language in Germany is going to be difficult, especially when it comes to face-to-face interviews. But you can carry out your investigation in German and then later translate it into English when you introduce your results in class.
On the other hand you can visit internationally operating companies in your home town where you might be able to conduct an investigation in the English language. You could enquire about the business connections, for example, the catchment areas of the workforce, suppliers and customers. In Geography classes it is customary to do a farm study. If you are lucky you might be able to do your farm study in English, but you should prepare your questionnaire in German, too. Class trips abroad offer a good opportunity to conduct surveys. Even in not-English-speaking countries there is a good chance that you find enough respondents for an investigation in English.
Recently it became more and more important to conduct surveys over the Internet. There are websites and software which help you to create the questionnaire and evaluate the results.

www.grafstat.de

Important tips for out-of-school research

- Plan your questions carefully. You must know how each question relates to your investigation.
- Plan your questions in a way that the answers can be easily recorded.
- Plan carefully when and where to administer your questionnaire and consider, in what way timing and location might influence your results.

Questionnaire on Gentrification

1. How long have you lived in this area?

☐ 0 - 2 years ☐ 3 – 5 years ☐ 6 – 10 years ☐ 11 years +

2. Where did you live before that?

Town:	District:

3. Is this your house? ☐ Yes ☐ No

4. What renovations have you made?

☐ Loft extension ☐ Conservatory ☐ Redecorating ☐ Other: ______

5. How many people live in your house? ___ 6. How many m² has it? _____

7. With regard to work, are you ...?

☐ Retired ☐ Part-time employed ☐ Full-time employed ☐ Unemployed

8. Where do/did you work?

Town:	District:

9. Do you have a car? ☐ Yes ☐ No 10. Do you have a second car? ☐ Yes ☐ No

Year: ______ Model: ______ Year: ______ Model: ______

11. Main mode of travel:

☐ Car/motorbike ☐ Bus ☐ Tube ☐ Bike ☐ Walk

12. Why did you choose to live here? (0 = *unimportant* ..... 5 = *very important*)

Prices		0	1	2	3	4	5
Near to:	communications	0	1	2	3	4	5
	workplace	0	1	2	3	4	5
	services	0	1	2	3	4	5
	good restaurants	0	1	2	3	4	5
Other (please state):		0	1	2	3	4	5

Make sure that you know exactly how to evaluate the questionnaire after the survey.

Few open questions – they take time and may be difficult to evaluate.

More closed questions – they are answered quickly and are easier to evaluate.

Make your questions short and simple, so that they are easy to understand.

Not more than 12 questions in door-to-door enquieries, even less in face-to-face interviews.

Use bipolar rating scales to assess opinions.

15306EX

M2 Questionnaire for a door-to-door enquiry

- Prepare enough questionnaire sheets.
- Use a clipboard as a support when you fill in the questionnaire.
- Be familiar with the technical terms you are using, if necessary, in German and in English.
- Being polite and dressing smartly will help you getting more respondents.
- Avoid offensive questions, for example, about age or income.
- Leave the more probing questions to the end.
- When you wish to interview specific people always arrange a meeting in advance.

(There are more useful hints next to the questionnaires in **M1** and **M2**.)

After you have carried out your investigation you should set up your findings in an interesting way, displaying diagrams and using presentation techniques which can be found in these skills pages.

USEFUL WORDS AND PHRASES

- Excuse me, I am doing a school geography project. Could I ask you one or two quick questions about ...
- We are going to conduct a survey about ...
- Our main aim is ...
- Could you please specify your answer.
- On a scale from 0 to 5, how important is it to ... / is ...
- Thank you very much for your help.
- I conducted an investigation into ...
- We chose face-to-face interviews to ask people about ...
- Our questionnaire was administered on a Wednesday afternoon between ...
- Nearly a quarter of the respondents believed/took the view that ...

Different Types of Numbers

1. Reading and writing numbers

.	**point** (for decimals, e.g. 2.54). See decimals.
,	Use a comma or a blank for groups of three numbers. (Do not read commas.) See below.
0	**zero**, "O", nought (UK)
1–10	In a text write **one** to **ten** as words.
>10	In a text write numbers **larger than** ten as numbers, e.g. "56", not "fifty-six".
123	a/one **hundred** (and) twenty-three
234	two **hundred** (and) thirty-four
5,678 (5 678)	five **thousand** six hundred (and) seventy-eight
93,000,000	ninety-three **million**
6,000,000,000	six **billion**

2. Different types of numbers

Ordinal numbers

1st	first	7th	seventh
2nd	second	8th	eighth
3rd	third	9th	ninth
4th	fourth	10th	tenth
5th	fifth	11th	eleventh
6th	sixth	12th	twelfth

Fractions

Fraction	We say ...	Decimal	Per cent
1/2	a/one half (of) / one out of two / one in two	0.5	50 %
1/3	a/one third (of) / one out of three / one in three	0.33	33 ⅓ %
1/4	a/one fourth (of) / one out of four / one in four	0.25	25 %
1/5	a/one fifth (of) / one out of five / every fifth	0.2	20 %
1/10	a/one tenth (of) / one out of ten / every tenth	0.1	10 %
1/20	a/one twentieth (of) / one out of twenty / one in twenty	0.05	5 %
1/100	a/one hundredth (of) / one out of a/one hundred	0.01	1 %

Decimals

0.5	(zero) point five
3.14 (π)	three point one four (pi)

Powers / Exponents

10^2	ten **squared** ten **to the power** of two ten **to the second**	**100**
10^3	ten **to the power of** three ten **to the third**	**1000**

The Earth is 93 x 10^6 miles from the sun.
(ninety-three times ten to the sixth) – 93 million –

3. Arithmetic

= Proportions

=	is, is **equal to, equals**	≠	is **unequal to**, is **not equa**
>	is **more**, is **larger than**	<	is **less**, is **fewer**, is **smaller t**
4:1	four **to** one	4 x	four **times as large as**

+ Addition → Sum (to add ... and ...)

12	+	4	=	16
Twelve	**plus/and**	four	equals/makes/is	sixtee

− Subtraction → Difference (to subtract ... from ...)

12	–	4	=	8
Twelve	**minus**	four	equals/makes/is	eigh

X Multiplication → Product (to multiply ... from ...)

12	x · *	4	=	48
Twelve	**times, (multiplied) by**	four	equals/makes/is	forty-ei

: Division → Quotient (to divide ... by ...)

12	: /	4	=	3
Twelve	**divided by**	four	equals/makes/is	thre

4. Geometry

Parallel lines

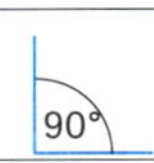

Right angle = 90°
A right angle has got ninety **degrees**.

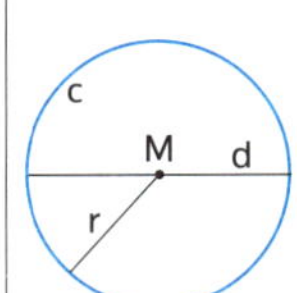

Circle
A circle is **round**. It has got a **centre** in the middle (**M**). The line from the centre to the circle wall is the **radius** (**r**). The line from on point of the circumference (**c**) through the centre to the opposite side is the **diameter**

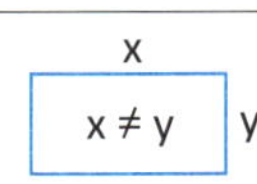

Rectangle or oblong
A rectangle has got two parallel equal side on two other equal sides at right angles.

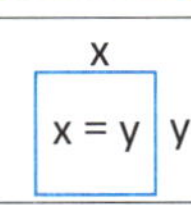

Square
A square is a rectangle with four equal sides.

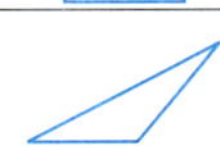

Triangle
A triangle is **triangular**. It has got three sic

3-d figures cube sphere pyramid cylinder cone

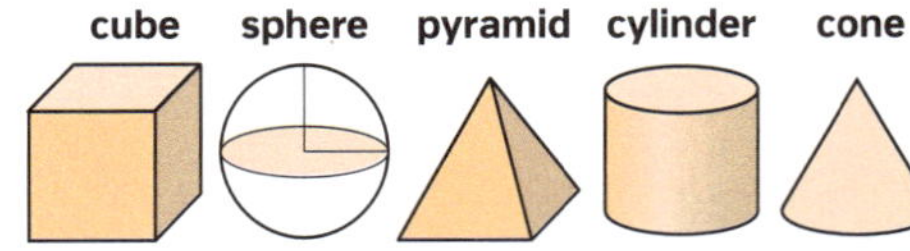

5. Time

1 minute	= 60 seconds
60 minutes	= 1 hour
24 hours	= 1 day
365 ¼ days	= 1 year
100 years	= 1 century

a.m. = from midnight t noon
p.m. = from noon to midnight

2007 = two tousand (an seven; twenty 'o' seve
2017 = twenty seventee

ifferent Types of Measurement

sides the standard metric system, the UK and the USA
o use a non-metric system. Basic measures are often
re than the metric ones: 1 inch > 1 cm, 1 mile > 1 km,
allon > 1 litre.

Precipitation – Height – Length – Distance

Non-metric		Metric	Uses
(inch)	=	2.54 cm	• amount of precipitation • map scales
foot)	12 **inches**	30.5 cm	• elevation above sea level • depth below sea level • width
yd ard)	36 **inches**/ 3 **feet**	0,914 m	middle distances (GB)
mile	5,280 **feet**/ 1760 **yd**	1.6 km	• distance

Metric	Non-metric
1 cm	**0.4 inches**
1 m	**39.4 inches**
1 km	**0.6 miles**

Speed

0 miles per hour = 160 km/h
0 kilometres per hour = 60 mph

Temperature

$^{o}C = (^{o}F - 32) / 1.8$
$^{o}F = 1.8 \times ^{o}C + 32$

he temperature in degrees Fahrenheit is equal to 1.8
mes the temperature in Celsius plus 32.

Celsius – Fahrenheit

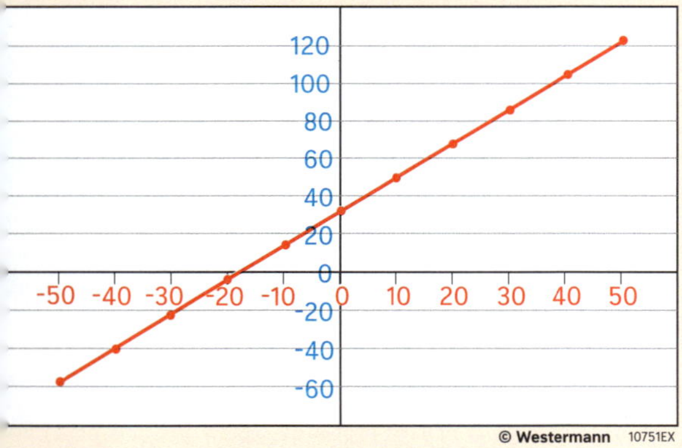

4. Area

1 sq inch = 6.45 sq cm
1 sq cm = 0.16 sq in
1 sq foot = 0.093 m^2
1 sq m = 10.76 sq ft
1 acre = 0.40 hectare
1 ha = 2.47 ac
1 sq mile = 2.59 sq km
1 sq km = 0.39 sq mi

5. Volume

Standard container size
20' x 8' x 8'6''
TEU (twenty-foot equivalent unit)
= 1360 cubic feet (cuft)
= 38.5 m^3

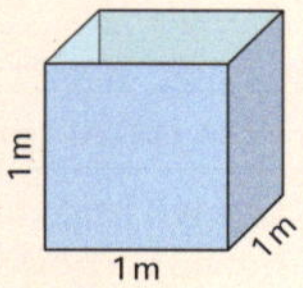

1 m x 1 m x 1 m = 1 m^3
1 metre by 1 metre by
1 metre = 1 cubic metre
= 1000 litres

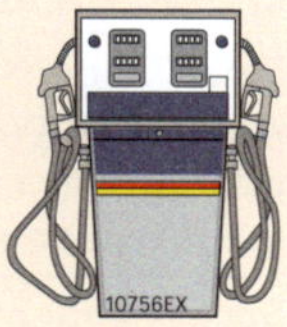

1 gallon is
4.6 litres (UK),
3.8 litres (USA).

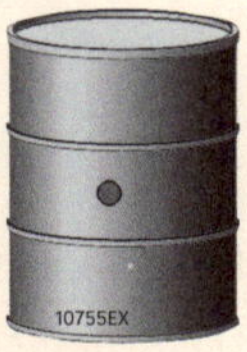

One barrel of oil contains 159 litres (42 US gallons).

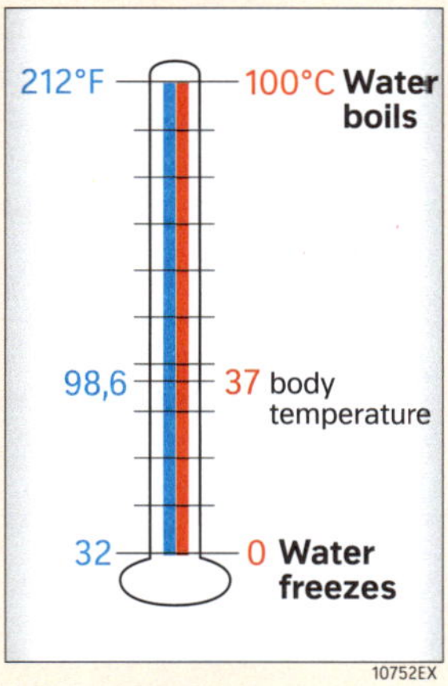

Internethinweis
zum
Informieren und
Umrechnen:

www.convertit.com

Instructions – Knowing What to Do

When you tackle a task you have to look out and observe instructions. They signal what to do when you solve the task.
In German, these instructions are called 'Operatoren'.
Most of the instructions of the Diercke Geography series are shown and explained here. These are the instructions you come across in your bilingual Geography lessons or in your exams and tests. The ones, however, that are not handled here are self-explaining, for example 'Write a text'. On the skills pages in this book you find helpful tips how to tackle most of these tasks.
We put the explanations in German, in order to make it as clear as possible what is expected from you in the particular task.

Instruction	**Operator**	**Was ist zu tun?**	**Beispiel**	**Hilfen**
Analyse	Analysiere	Die wichtigsten Merkmale herausarbeiten. Dann im Detail – gegebenenfalls kritisch – untersuchen.	Analyse the Rhine-Main region as a potential location for a company in the service industry.	Eigene Stellungnahme möglich.
Assess	Bewerte	Einen Gegenstand an erkennbaren Wertkategorien oder an bekannten Beurteilungskriterien messen.	Assess the measures taken to improve the quality of life for people living in shrinking cities.	Eigenen Standpunkt deutlich machen.
Comment on	Kommentiere	Sachbezogen beurteilen und abschließend bewerten.	Comment on the role of renewables in the world energy supply.	›S. 35
Compare	Vergleiche	Vergleichbares nebeneinander stellen und die Gemeinsamkeiten und Unterschiede herausarbeiten.	Compare Germany's population structure of 1910 with that of today.	›S. 36
Contrast	Nenne die Unterschiede zwischen	Die Unterschiede Punkt für Punkt gegenüberstellen.	Contrast the challenges Tanzania and Switzerland are facing due to their age structures and their stage of development.	›S. 36
Create / Make a diagram / mind map / concept map / poster / ...	Erstelle ein(e) Diagramm / Mindmap / Concept Map / Poster / ...	Ein Medium in ein anderes überführen. Gegebenenfalls zusätzliche Informationen sammeln.	Create a flow chart about the effect of a drought on hydropower.	Viele Skills-Seiten (›S. 16 - 29) zeigen, wie man es macht.
Describe	Beschreibe	Einen Tatbestand (Tabelle, Graphik) durch umfassende Angaben verständlich machen.	Describe the formation of crude oil and its extraction.	Es sind keine Erklärungen erforderlich.

Instruction	Operator	Was ist zu tun?	Beispiel	Hilfen
Discuss	Erörtere / Diskutiere	Das Für und Wider bzw. die positiven und negativen Seiten eines Sachverhaltes darstellen und zu einem eigenen Urteil kommen.	Discuss the use of solar power in LEDCs and MEDCs.	›S. 37
Evaluate	Werte ... aus	Daten, Einzelergebnisse oder andere Elemente in einen Zusammenhang stellen und ggf. zu einer Gesamtaussage zusammenführen.	Evaluate the changes of Detroit's CBD.	Entscheide, wie wichtig, sinnvoll, wertvoll oder wirkungsvoll etwas ist, und begründe deine Entscheidungen.
Explain	Erkläre	Durch Darstellen der Ursachen bzw. Zusammenhänge das Verstehen eines Sachverhaltes ermöglichen.	Explain how the terms of trade of Ghana and Nigeria have changed since 1995.	›S. 34
Give reasons	Begründe	Ursachen für einen Sachverhalt oder eine Einstellung benennen.	Give reasons for the differences in worldwide population distribution.	Setzt das Nennen (Name) und Erklären (Explain) voraus. Ohne eigene Stellungnahme.
Illustrate	Erläutere	Einen Sachverhalt veranschaulichend darstellen und durch zusätzliche Informationen verständlich machen.	Illustrate the role of women in the Indian plastic recycling industry.	Kombination von Beschreiben (Describe) und Erklären (Explain).
List / Make a list	Erstelle eine Liste	Eine knappe, geordnete Zusammenstellung von Fakten in Stichpunkten.	List the causes of soil erosion in the Dust Bowl.	Punkte mit Spiegelstrichen gliedern.
Locate	Verorte	Einen Standort bestimmen (Gradnetz, Atlas).	Locate the Fehmarn Belt.	›S. 6 - 7 Geographische Angaben aus der Umgebung als Erläuterung möglich.
Name	Nenne	Informationen unkommentiert wiedergeben.	Name characteristics of a global city.	Einzelne Punkte, mit Überleitungen verbinden.
Prepare (for) a discussion	Bereite (dich für) eine Diskussion vor	Fakten und unterschiedliche Standpunkte zu einem Sachverhalt (schriftlich) zusammentragen.	Prepare for a discussion in class about your individual ecological footprints.	›S. 44 - 45
Prepare and give a talk / speech / presentation	Bereite einen Vortrag / eine Präsentation vor und halte ihn / sie.		Prepare a speech, in German, about a nuclear accident.	›S. 40 - 41
Research (on)	Recherchiere	Im Internet und anderen Quellen Informationen finden und zusammenstellen.	Research which articles of your daily use are made from oil products.	Arbeite zielgerichtet und vergleiche mehrere Quellen zum Thema. Kennzeichne Zitate.
Summarise / Sum up / Outline	Fasse zusammen	In vollständigen Sätzen den Inhalt eines Textes und anderer Materialien knapp wiedergeben. Die Struktur nachvollziehen.	Sum up the forms of production presented on this page.	Vermeide es, Textpassagen zu zitieren.

Glossary

A

absorption [əb'sɔ:pʃn]
Absorption – Uptake of CO_2 by → biomass through photosynthesis.

Agenda 21 [ə'dʒendə 'twentiwʌn]
Agenda 21 – United Nation action plan to be taken globally, nationally and locally with regard to → sustainable development.

agglomeration [əˌglɒmə'reɪʃn]
Ballungsraum, Agglomeration – Major → urban area, often resulting from the merging of cities caused by the process of → suburbanisation.

agribusiness ['ægrɪˌbɪznəs]
Agrarindustrie – Farming on a large scale, often including the marketing of farm produce.

agriculture ['ægrɪkʌltʃə(r)]
Landwirtschaft – Cultivating land, raising crops, feeding, breeding, and raising → livestock.

agroforestry [agrəʊ'fɒrɪstri]
Agroforstwirtschaft – Combination of agricultural activities and → forestry.

altitudinal zone [altɪ'tju:dɪn(ə)l zəʊn]
Höhenstufe – Typical pattern of → vegetation that changes with height due to different climatic conditions.

ancillary industry [æn'sɪləri 'ɪndəstri]
Zulieferindustrie – Manufactures parts and components to be used in the production process; e.g. a company which produces engines needed for the production of a car.

Antarctic Circle [æn'tɑ:(r)ktɪk 'sɜ:(r)k(ə)l]
südlicher Polarkreis – Line of latitude at 66.5° S. Between there and the South Pole lies the → Polar Zone.

aquifer ['ækwɪfə(r)]
Grundwasserleiter – Water-bearing layer of rock or other material above an impermeable layer which can absorb and store → groundwater.

arable farming ['ærəb(ə)l 'fɑ:(r)mɪŋ]
Ackerbau – Growing of crops on farms.

Arctic Circle ['ɑ:(r)ktɪk 'sɜ:(r)k(ə)l]
nördlicher Polarkreis – Line of latitude at 66.5° N. Between there and the North Pole lies the → Polar Zone.

artesian well [ɑ:(r)ˌti:ziən 'wel]
Artesischer Brunnen – Hole in the ground from which water rises under pressure to the surface from a permeable layer overlaid by impermeable rock.

assembly plant [ə'sembli plɑ:nt]
Montagewerk – Factory in which parts, e.g. of a car, are put together.

asthenosphere [əs'θɛnəsfɪə]
Asthenosphäre – Viscous part of the → upper mantle directly below the solid lithosphere. The → plates of the → lithosphere float on this layer.

autonomous territory [ɔ:'tɒnəməs 'tɛrɪt(ə)ri]
autonomes Gebiet – Region which has a certain degree of political autonomy or is free from an external authority.

B

balance of trade [ˌbæləns əv 'treɪd]
Handelsbilanz – Difference in value between → imports and → exports.

basin ['beɪs(ə)n]
Becken – Depression or low area on the land's surface.

biocapacity ['baɪəʊkə'pæsəti]
Biokapazität – Ecosystems' ability to produce → biomass used by mankind and to absorb waste material generated by them.

biodiversity [ˌbʌɪə(ʊ)dʌɪ'və:sɪti]
Biodiversität – The number and variety of plants and animals found within an → ecosystem.

bioenergy [ˌbaɪəʊ'enədʒi]
Bioenergie – Energy contained in plants, animals, bacteria and fungi and their residues.

biofuel ['baɪəʊfju:əl]
Biokraftstoff – Gaseous, liquid, or solid substance composed of or produced from biological raw materials that is used as a → fuel.

biogas ['baɪəʊgæs]
Biogas – Gas produced by the action of bacterial decomposition of sewage, manure, garbage, or plant crops and used as a → fuel.

biomass ['baɪəʊmæs]
Biomasse – Total mass of all living things.

birth rate ['bɜ:θ reɪt]
Geburtenrate – Figure, given in per mil (‰), describing the number of people born per thousand inhabitants in a certain area within one year.

bituminous coal [bɪ'tju:mɪnəs kəʊl]
Steinkohle – Usually black coal, used primarily as → fuel in electricity generation, also used for heat and power applications in manufacturing.

blizzard ['blɪzəd]
Blizzard, Schneesturm – Severe snowstorm with high winds in North America.

body of water ['bɒdi ɒv 'wɔ:tə(r)]
Gewässer – Accumulations of water on the surface of the → Earth, such as → oceans, seas, lakes, ponds, rivers, streams, and canals.

border ['bɔ:(r)də(r)]
Grenze – Line separating two systems, either man-made (e.g. → states) or natural (e.g. plates).

boreal forest [b'əʊrɪəl 'fɒrɪst]
borealer Nadelwald – → Vegetation zone consisting of forests with trees having mostly needle-shaped leaves. In the → Northern Hemisphere, it lies south of the → tundra region. It has short, warm summers, and long, cold winters with plenty of snow.

brain drain ['breɪn dreɪn]
Brain Drain – → Emigration of highly skilled and qualified people regarded as harmful to a country's → economy or society.

brain gain ['breɪn geɪn]
Brain Gain – → Immigration of highly trained or qualified people, regarded as beneficial to a country's → economy or society.

business district ['bɪznəs 'dɪstrɪkt]
Geschäftsviertel – Part of a city or town where there are many businesses.

C

canopy ['kanəpi]
Kronendach – Roof-like upper layer in a forest, e. g. the → tropical rainforest, formed by tree crowns.

capital ['kæpɪt(ə)l]
Hauptstadt – City with a seat of government.

car manufacturer [kɑ:(r) ˌmænju'fæktʃərə(r)]
Autohersteller – Company which produces cars.

carbon cycle ['kɑ:bən saɪkl]
Kohlenstoffkreislauf – The exchange of carbon between pedosphere, geosphere, hydrosphere, and atmosphere.

carbon emission trading ['kɑ:bən i'mɪʃn 'treɪdɪŋ]
Emissionshandel mit Kohlendioxid – Government-mandated instrument to reduce → greenhouse gas → emissions.

carbon footprint [ˌkɑ:bən 'fʊtprɪnt]
CO_2-Fußabdruck – Total amount of greenhouse gases produced to directly and indirectly support human activities, usually expressed in equivalent tons of carbon dioxide.

cash crop [kæʃ krɒp]
Marktfrucht – Crop which is grown to be sold.

caste system [kɑ:st 'sɪstəm]
Kastenwesen – System of social grouping in India in which the social status of a person is determined by birth.

central business district, CBD ['sentrəl 'bɪznəs 'dɪstrɪkt]
Hauptgeschäftsviertel – Area of a town or a city which is the centre of → administration, shopping, business, and entertainment.

circular economy ['sɜ:kjələ(r) ɪ'kɒnəmi]
Kreislaufwirtschaft – Economic system in which resource input and waste, → emission, and energy leakage are minimised through long-lasting design, sharing, repair, reuse, refurbishing, re-manufacturing, and recycling.

city ['sɪti]
(Groß-)Stadt – Concentration of people and buildings with an inner central business district and the → outskirts where most people live.

clean coal technology [kli:n kəʊl tek'nɒlədʒi]
saubere Kohleumwandlungstechnologie – Reduces harmful environmental effects like CO_2 and particulates → emissions by using multiple filter, conversion, and storage technologies.

climate ['klaɪmət]
Klima – Average weather conditions of a certain region over at least 30 years.

climate agreement ['klaɪmət ə'gri:mənt]
Klimaabkommen – Binding agreement between nations of the world on tackling → climate change by cutting → greenhouse gas → emissions.

climate change ['klaɪmət tʃeɪndʒ]
Klimawandel – Change in global or regional climates over a long period of time; the variation of climatic elements over time.

climate protection ['klaɪmət prə'tekʃn]
Klimaschutz – Agreement to protect the global climate; mainly the stabilisation of → greenhouse gas concentrations in the atmosphere to prevent → global warming.

climate zone ['klaɪ'mæt zəʊn]
Klimazone – Area with a consistent → climate. There are five basic → climatic zones which mainly depend on radiation: → Polar, Subpolar, Temperate, Subtropical, and Tropical Zone.

coal mining crisis [kəʊl 'maɪnɪŋ 'kraɪsɪs]
Kohlenkrise – Less demand and low prices led to a decline of the coal mining industry that started in the 1950s and effected the → old industrial regions of Europe.

coastal plain ['kəʊstl pleɪn]
Küstenebene – Flat land near the sea.

coastline ['kəʊs(t)lʌɪn]
Küstenlinie – Shape, outline, or boundary of a coast.

collision margin [kə'lɪʒ(ə)n 'mɑ:dʒɪn]
Kollidierende Plattengrenze / Erdkruste aufschiebender Rand – Area of the → Earth's crust where two → plates of → continental crust move towards each other, creating fold mountains.

colonisation [ˌkɒlənaɪ'zeɪʃ(ə)n]
Kolonialisierung – Political, economic and legal domination of one country over another one, often beginning with establishment of trading posts and settlements or cultivation of land.

combined heat and power (CHP)
[kəm'baɪnd hi:t ənd 'paʊə(r)]
Kraft-Wärme-Kopplung – Combines the production of usable heat and electricity in one single, efficient process.

commodity [kə'mɒdəti]
Rohstoff, Handelsgut – Unprocessed or partially processed good which can be traded for the production of other goods.

Common Agricultural Policy (CAP)
['kɒmən ˌægrɪ'kʌltʃərəl 'pɒləsi]
Gemeinsame Agrarpolitik (CAP) – Agricultural policy of the → EU.

component supplier [kəm'pəʊnənt sə'plaɪə(r)]
Zulieferer / Zulieferfirma – Company producing components for the final assembly in another company.

concentrated solar power (CSP)
['kɒnsntreɪtɪd 'səʊlə(r) 'paʊə(r)]
solarthermische Energie – Mirrors or lenses are used to focus a large area of sunlight onto a small area.

condensation [ˌkɒndenˈseɪʃ(ə)n]
Kondensation – Transition of water vapour into water droplets whereby heat is released.

coniferous forest [kəˈnɪf(ə)rəs ˈfɒrɪst]
Nadelwald – Forest with trees having needle-shaped leaves.

conservative margin
[kənˈsɜː(r)vətɪv ˈmɑː(r)dʒɪn]
Konservative Plattengrenze / Erdkruste bewahrender Rand – Area of the → Earth's crust where two → plates slide past each other.

constructive margin [kənˈstrʌktɪv ˈmɑːdʒɪn]
Konstruktive Plattengrenze / Erdkruste bildender Rand – Area of the → Earth's crust where two → plates move apart creating gaps, continuously filled with → magma or lava.

container vessel [kənˈteɪnə(r) ˈvesl]
Containerschiff – Type of ship that is built to be used in global trade to ship containers.

continent [ˈkɒntɪnənt]
Kontinent – Very large land area surrounded by → oceans.

continental climate [ˌkɒntɪˈnent(ə)l ˈklaɪmət],
Kontinentalklima – Climate with high temperature range between warm summers and cold winters.

continental crust [ˌkɒntɪˈnent(ə)l krʌst]
kontinentale Kruste – Thicker but lighter part of the → Earth's crust.

continental drift [ˌkɒntɪˈnent(ə)l drɪft]
Kontinentaldrift – Alfred Wegener's theory of the movement of → continents.

contract farming [ˈkɒntrækt ˈfɑː(r)mɪŋ]
Vertragsanbau – Employment of specialised firms for specific farming activities on a previously agreed basis.

convection currents [kənˈvekʃ(ə)n ˈkʌrənts]
Konvektionsströme – Upward or downward movement of → magma within the viscous part of the → upper mantle caused by heat.

conventional gas [kənˈvenʃənl gæs]
Gas aus konventionellen Ressourcen – → Natural gas occurring in porous and permeable rock is either in the gaseous phase or dissolved in → crude oil, which can be extracted by normal production practices.

coral reef [ˈkɒr(ə)l riːf]
Korallenriff – Ecosystems located underwater which are built by corals. They offer shelter and food for many other species.

core city [kɔː(r) ˈsɪti]
Kernstadt – City in its administrative boundaries.

crop [krɒp]
Feldfrucht / Nutzpflanze – Cultivated plant, especially a cereal, fruit, or vegetable.

crude oil [kruːd ɔɪl]
Rohöl – Petroleum as it comes out of the → oil well, before refining.

cultivation [kʌltɪˈveɪʃn]
Anbau – → Arable farming.

cultivation limit [kʌltɪˈveɪʃn ˈlɪmɪt]
Anbaugrenze – Limit, until which → arable farming is possible. It can be determined by height, cold or drought.

D

dairy farming [ˈdeɪri ˈfɑː(r)mɪŋ]
Milchviehhaltung – Rearing of cows for milk production.

death rate [ˈdeθ reɪt]
Sterberate – Figure, given in per mil (‰), describing the number of deaths per thousand inhabitants in a certain area within one year.

decarbonisation [ˌdiːˌkɑːbənaɪˈzeɪʃn]
Entkarbonisierung – Reduction of the use of → fossil fuels which all produce carbon dioxide.

decentralisation [ˌdiːˌsentrəlaɪˈzeɪʃn]
Dezentralisierung – Dispersion of a concentration of political power, industry, or population.

deciduous forest [dɪˈsɪdjuəs ˈfɒrɪst]
Laubwald – Forest with trees which shed leaves in winter because of the lack of water in the frozen soil.

deforestation [diːˌfɒrɪˈsteɪʃ(ə)n]
Abholzung, Entwaldung – Clearing of forest to provide firewood, building materials or to use the land for economic activities.

delta [ˈdeltə]
Delta – → River mouth at a → coastal plain or a lake made up of channels and islands, often shaped like a fan stretching out into the sea.

demographic change [ˌdeməˈgræfɪk tʃeɪndʒ]
demographischer Wandel – Shift in population structure that has emerged in many industrialised nations since the 1970s, chiefly as a result of three major developments:
declining → birth rates, longer → life expectancy and increasing → immigration.

demographic transition [ˌdeməˈgræfɪk trænˈzɪʃn]
demographischer Übergang – Theory of the passage from high birth and death rates to low birth and death rates in any population during its economic development.

Demographic Transition Model (DTM)
[ˌdeməˈgræfɪk trænˈzɪʃn ˈmɒdl]
Modell des demographischen Übergangs – Pattern describing the → demographic transition.

dependent [dɪˈpendənt]
Abhängige(r) – Non-working population below 15 and above 65 years of age.

deposit [dɪˈpɒzɪt]
Lagerstätte – An area in the Earth's crust where → resources are found in a high natural concentration.

desalinisation [ˌdiːsælɪˈneɪʃ(ə)n]
Entsalzung – Process which removes salt and other minerals from water, mainly to produce → fresh water.

desert [ˈdezə(r)t]
Wüste – Dry zone covered with rock, gravel or sand, but only sparse → vegetation.

desertification [dɪˌzɜː(r)tɪfɪˈkeɪʃ(ə)n]
Desertifikation – Man-made or natural destruction of → vegetation, leading to desert-like landscapes.

destructive margin [dɪˈstrʌktɪv ˈmɑːdʒɪn]
Destruktive Plattengrenze / Erdkruste zerstörender Rand – Area of the → Earth's crust where heavier → oceanic crust moves under lighter → continental crust.

development [dɪˈveləpmənt]
Entwicklung – Process in which a region becomes more advanced economically, environmentally and/or socially.

digital divide [ˈdɪdʒɪtl dɪˈvaɪd]
digitale Kluft – Differences in accessibility and usage of information and communication technology in different regions of the Earth.

disparity [dɪˈspærəti]
Unterschied, Disparität – Inequality in wellbeing, especially in economic wealth.

district heating [ˈdɪstrɪkt ˈhiːtɪŋ]
Fernwärme – System for distributing heat generated in a centralised location for residential and commercial space and water heating.

downtown [ˌdaʊnˈtaʊn]
Innenstadt, Stadtzentrum – American for: Central part of a city, especially its main commercial area.

drinking water [drɪŋkɪŋ ˈwɔːtə]
Trinkwasser – → Freshwater safe enough for humans to drink.

drip irrigation [drɪp ˌɪrɪˈgeɪʃ(ə)n]
Tröpfchenbewässerung – System of → crop → irrigation involving the controlled delivery of water directly to individual plants through a network of pipes.

drought [draʊt]
Dürre – Long period of time when an area does not get enough water.

dry season [drʌɪ ˈsiːz(ə)n]
Trockenzeit – Yearly period of low rainfall in tropical regions.

E

Earth [ɜː(r)θ]
Erde – → Planet on which we live which is the third planet in the → solar system.

earthquake [ˈɜː(r)θˌkweɪk]
Erdbeben – Vibration of the → Earth‘s crust caused by movements of → plates, by → volcanic eruptions, or by → fracking.

Earth's axis [ɜː(r)θs ˈæksɪs]
Erdachse – Imaginary line running from the → North Pole to the → South Pole around which the → Earth seems to spin.

Earth's core [ɜː(r)θ kɔː(r)]
Erdkern – Innermost part of the Earth, mainly consisting of the two metals nickel and iron.

Earth's crust [ɜː(r)θ krʌst]
Erdkruste – Uppermost layer of the Earth, forming the solid surface of the Earth.

Earth's mantle [ɜː(r)θ ˈmant(ə)l]
Erdmantel – Layer between the → Earth's crust and the → Earth's core.

ecological footprint [ˌiːkəˈlɒdʒɪkl ˈfʊtprɪnt]
Ökologischer Fußabdruck – Area of productive land and water required to produce the resources that a population consumes and assimilate the wastes that this population produces.

ecological overshoot [ˌiːkəˈlɒdʒɪkl ˌəʊvəˈʃuːt]
Überschreiten der ökologischen Belastbarkeitsgrenze –
Moment when humanity's annual demand on the natural world has exceeded what the Earth can renew in a year.

economic crisis [ˌiːkəˈnɒmɪk ˈkraɪsɪs]
Wirtschaftskrise – Situation in which the → economy of a country experiences a downturn connected with a falling → GDP.

economic power [ˌiːkəˈnɒmɪk ˈpaʊə(r)]
Wirtschaftskraft – Ability to influence because of being able to produce, buy, or sell a lot.

economic sector [ˌiːkəˈnɒmɪk ˈsektə(r)]
Wirtschaftssektor – Category of activities in the production of goods.

economy [ɪˈkɒnəmi]
Wirtschaft – Sum of all activities of the → primary, → secondary, and → tertiary sectors (industries).

ecosystem [ˈiːkəʊˌsɪstəm]
Ökosystem – System formed by the interaction of all living organisms with each other and with the physical and chemical factors of the → environment in which they live.

electric mobility [ɪˈlektrɪk məʊˈbɪləti]
Elektromobilität – One concept of the → Energiewende that vehicles primarily use electric energy.

emigration [ˌemɪˈgreɪʃn]
Auswanderung – Movement of people out of a country.

emission [iˈmɪʃn]
Emission, Ausstoß – Sending out of substances.

Energiewende
Energiewende – Part of a German policy of → energy transition which aims to reduce the use of → fossil fuels and abolish nuclear fuels in order to create an energy system based on at least 60 % → renewable energy by 2050.

energy conversion [ˈenədʒi kənˈvɜːʃn]
Energieumwandlung – Transition of one form of energy to another; always connected with an energy loss.

energy crop [ˈenədʒi krɒp]
Energiepflanze – Food → crop and non-food crop grown for its → fuel value.

energy efficiency [ˈenədʒi ɪˈfɪʃnsi]
Energieeffizienz – Percentage of total energy input that is useful and not wasted as useless heat.

energy transition [ˈɛnədʒi trænˈzɪʃn]
Energiewende – Transition towards sustainable energy, decentralised → renewable energy, and → energy efficiency.

environment [ɪnˈvaɪrənmənt]
Umwelt – All living and non-living things with their interactions that affect an organism or a group of organisms.

environmental problem [ɪnˌvaɪrənˈment(ə)l ˈprɒbləm]
Umweltproblem, Umweltschaden – Damage in the → environment caused by human activities.

epicentre [ˈepɪˌsentə(r)]
Epizentrum – Location with the strongest vibrations on the Earth's surface directly above the place from where an → earthquake started (→ hypocentre).

Equator [ɪˈkweɪtə(r)]
Äquator – Longest line of latitude with an equal distance from the two poles, dividing the → Northern and → Southern Hemisphere.

erg [ˈɜː(r)g]
Sandwüste – Sandy → desert landscape with sand dunes, which makes up about 20 per cent of the Sahara Desert.

erosion [ɪˈrəʊʒ(ə)n]
Bodenabtrag (Erosion) – Destructive effect on the Earth's surface caused by running water, wind, moving ice, and gravity, leading to the transport of material.

estuary [ˈestjuəri]
Ästuar – → River mouth which is influenced by the → tides and does not stretch out into the sea.

ethnic group [ˈɛθnɪk gruːp]
Ethnische Gruppe – Group of people who identify with each other based on a common social, cultural or national experience.

European Central Bank (ECB) [ˌjʊərəˈpiːən ˌsentrəl ˈbæŋk]
Europäische Zentralbank – Bank responsible for the monetary system of the → European Union and the euro.

European Community (EC) [ˌjʊərəˈpiːən kəˈmjuːnəti]
Europäische Gemeinschaft (EG) – Organisation (1993 – 2009) to unify the European economies and now part of the → European Union.

European Union (EU) [ˌjʊərəˈpiːən ˈjuːniən]
Europäische Union – Economic and political organisation that many European countries belong to.

Euroscepticism [ˈjʊərəʊ ˈskeptɪsɪzəm]
EU-Skepsis – Very critical view of the → European Union.

evaporation [ɪˌvæpəˈreɪʃ(ə)n]
Verdunstung – Transition of liquid water into water vapour for which energy is needed.

export [ɪkˈspɔːt]
Ausfuhr (Export) – Process of selling goods or services to other countries as well as their financial value.

export good [ɪkˈspɔːt gʊd]
Exportgut – Product that is sold and sent from one country to another.

extraction [ɪkˈstrækʃn]
Förderung – Process by which → crude oil or → natural gas is produced.

F

farming [ˈfɑːmɪŋ]
Landwirtschaft – Use of land for growing → crops or breeding and keeping → livestock.

farming type [ˈfɑː(r)mɪŋ taɪp]
landwirtschaftlicher Betriebstyp – Classification of farms, depending on production (e.g. → dairy farming), ownership (e.g. family farm) or market orientation (e.g. → subsistence farming).

favela [fəˈvelə]
Favela – Portuguese word for unplanned settlements with poor living conditions; located on the → outskirts of large cities as well as within them.

final assembly [ˈfaɪn(ə)l əˈsembli]
Endfertigung – Last step in a production line in which all components are put together.

fish farm [ˈfɪʃ fɑːm]
Fischzuchtbetrieb – Form of aquaculture to raise fish commercially in tanks or enclosures, usually for food.

fish stock [ˈfɪʃ stɒks]
Fischbestand – Population or total mass of a fishery resource.

flood protection [flʌd prəˈtekʃn]
Hochwasserschutz – Methods used to reduce or prevent the effects of flood waters or high water levels.

floodplain [flʌdpleɪn]
Überschwemmungsgebiet, Flussaue – Flat land between a river and the slopes of a → valley which is flooded occasionally or regularly.

food security [fuːd sɪˈkjʊərəti]
Ernährungssicherheit – Supply of food and people's access to it.

footloose industry [ˈfʊtˌluːs ˈɪndəstri]
Footloose Industry, Ungebundene Industrie – Industry where the location is not influenced by access to → raw materials or markets.

forecast [ˈfɔː(r)kɑːst]
Vorhersage – A description of what something (e.g. the weather) will be like in the future.

forestry [ˈfɒrɪstri]
Forstwirtschaft – Planting, managing, and harvesting large areas of trees.

fossil fuel [ˈfɒsl fjuːəl]
fossiler Brennstoff – → Fuel (coal, oil, or → natural gas) formed in the → Earth's crust from plant and animal remains.

fracking [ˈfrækɪŋ]
Fracking – Method of extracting oil or gas from a layer of rock by horizontal drilling and creating cracks.

fragmentation [ˌfrægmenˈteɪʃn]
Fragmentierung – Increasing social or economic inequalities in the process of → globalisation.

free market economy [friː ˈmɑː(r)kət ɪˈkɒnəmi]
freie Marktwirtschaft – Economic system with a government not interfering in business activities.

free trade [friː treɪd]
Freihandel – Free exchange of goods between countries without restrictions such as taxes on → imports / exports.

GLOSSARY

free trade zone [ˌfri: treɪd zəʊn]
Freihandelszone – Special area within a country where foreign companies can import materials, manufacture goods, export products, etc. without being limited by the rules and taxes of the country.

fresh water [freʃ ˈwɔ:tə(r)]
Süßwasser – The naturally occurring water on Earth's surface in ice sheets, ice caps, glaciers, icebergs, bogs, ponds, lakes, rivers and streams, and groundwater in aquifers and underground streams.

fuel [fju:əl]
Brennstoff – An energy source for engines, → power stations, or reactors.

fuel cell [fju:əl sel]
Brennstoffzelle – Device that produces electricity through a chemical reaction of a source fuel directly from the oxidation of a → fuel, for example, hydrogen by oxygen.

G

gas terminal [gæs ˈtɜ:(r)mɪn(ə)l]
Gasterminal – → Oil terminal.

gated community [ˈgeɪtɪd kəˈmju:nəti]
bewachte Wohnanlage – Residential area with controlled entrance and sometimes fences and walls, mostly inhabited by rich and/or elderly people.

general fertility rate [ˈdʒenrəl fəˈtɪləti reɪt]
Fruchtbarkeitsrate – Figure, usually given in per mil (‰), showing the average number of children born to a woman in her lifetime in a certain area.

genetically modified (GM) crops [dʒəˈnɛtɪkli ˈmɒdɪfʌɪ (dʒi: ɛm) krɒps]
gentechnisch veränderte Nutzpflanzen – Crops whose genetic characteristics have been altered using the techniques of genetic engineering.

gentrification [ˌdʒentrɪfɪˈkeɪʃn]
Gentrifizierung – Movement of wealthy people mostly into formerly neglected parts of a city after the renovation and restoration of buildings and flats; mostly combined with the outward → migration of the former inhabitants who cannot afford to live there any longer.

geothermal energy [ˌdʒi:əʊˈθɜ:ml ˈenədʒi]
Geothermalenergie, Erdwärmeenergie – Energy produced by extracting the Earth's internal heat originating in its core.

ghetto [ˈgetəʊ]
Ghetto – Highly segregated area in a city where a minority, often an → ethnic group, lives in poor housing conditions; originally the residential concentration of Jews.

global city [ˈgləʊbl ˈsɪti]
Global City, Globalstadt – Cities which are the most important political, economic, cultural and financial centres of the world.

global grid [ˈgləʊb(ə)l grɪd]
Gradnetz der Erde – Net of virtual → lines of latitude and longitude with which you can position everything on → Earth.

global player [ˈgləʊbl ˈpleɪə(r)]
Global Player, Weltkonzern – Corporation or enterprise that manages production or delivers services on a worldwide scale (multinational corporation).

global sourcing [ˈgləʊbl sɔ:sɪŋ]
Global Sourcing, weltweiter Einkauf – Economic strategy with the aim of finding the most cost efficient location worldwide to manufacture a product.

global warming [ˌgləʊbl ˈwɔ:mɪŋ]
Erderwärmung – Increase in the Earth's average atmospheric temperature that causes changes in climate and likely is the result of the man-made → greenhouse effect.

globalisation [ˌgləʊbəlaɪˈzeɪʃn]
Globalisierung – Process by which national and regional economies, societies, and cultures have become integrated through the global network of trade, communication, transportation, and → migration.

globe [gləʊb]
Globus – → Earth as a whole. Also: A small 3-D model of it.

government district [ˌgʌvnˈment ˈdɪstrɪkt]
Regierungsviertel – Part of the city mainly used by state institutions like parliament, law courts, ministries, embassies, etc.

GPS (Global Positioning System)
[ˌdʒi: pi: ˈes, (ˈgləʊb(ə)l pəˈzɪʃ(ə)nɪŋ ˈsɪstəm)]
Globales Navigationssystem – Device to show the exact location of a person or thing on Earth through signals from satellites.

greenhouse effect [ˈgri:nhaʊs ɪfekt]
Treibhauseffekt, Glashauseffekt – Warming of the atmosphere through gases like carbon dioxide, methane and water vapour which allow shortwave radiation (light) to pass through the atmosphere but absorb some of the longwave radiation (heat) emitted from the Earth's surface.

greenhouse gas [ˌgri:nhaʊs ˈgæs]
Treibhausgas – Gas in the atmosphere, mainly carbon dioxide, methane, ozone, and water vapour, that absorbs solar radiation, emits heat radiation and is responsible for the → greenhouse effect.

gross domestic product (GDP) [ˌgrəʊs dəˌmestɪk ˈprɒdʌkt]
Bruttoinlandsprodukt (BIP) – Value of all finished goods and services which are produced or rendered in a country within one year.

gross national product (GNP) [ˌgrəʊs ˈnæʃnəl ˈprɒdʌkt]
Bruttonationalprodukt (BNP) – Total value of all the goods and services produced by a country in one year, including the total income from foreign countries.

groundwater [ˈgraʊndˌwɔ:tə(r)]
Grundwasser – Water which is stored in an → aquifer.

groundwater table [ˈgraʊndˌwɔ:tə(r) ˈteɪb(ə)l]
Grundwasserspiegel – Upper level of the → groundwater contained in an → aquifer. It changes due to the input and output of water.

growth industry [grəʊθ ˈɪndəstri]
Wachstumsindustrie – Sector of the economy experiencing a higher-than-average growth rate.

H

habitat [ˈhabɪtat]
Lebensraum – Natural → environment where a particular type of animal or plant is usually found.

hamada [ˈhæmædæ]
Schuttwüste – → Desert landscape covered with rock and stones, which makes up 70 per cent of the Sahara Desert.

headquarters [ˌhedˈkwɔːtəz]
Hauptsitz, Hauptniederlassung – Main offices of a company, organisation, or institution.

heat pump [hiːt pʌmp]
Wärmepumpe – Device that takes heat from one location and moves it to another location; used either for the purpose of heating or cooling.

heavy industry [ˈhevi ˈɪndəstri]
Schwerindustrie – Industry that uses large machinery to produce raw materials for other industries or bulky goods, often causing disruption to the environment.

hemisphere [ˈhemɪˌsfɪə(r)]
Halbkugel – One half of the → Earth, either divided along the → Equator (→ Northern and Southern Hemisphere) or by the → Prime Meridian and 180° meridian (Eastern and Western Hemisphere).

high mountain range [hʌɪ ˈmaʊntɪn reɪn(d)ʒ]
Hochgebirge – A unit of several mountains, usually above 1500 m.

high-tech industry [haɪ tek ˈɪndəstri]
Hochtechnologieindustrie – Industry using advanced skills and machines, especially electronic ones.

hill farming [hɪl ˈfaː(r)mɪŋ]
Berglandwirtschaft – Farming in a → mountainous area where the prevailing grassland is mainly used for cattle or sheep grazing.

holiday region [ˈhɒlɪdeɪ ˈriːdʒ(ə)n]
Urlaubsregion – Area which is popular for tourists to spend their holidays.

hub [hʌb]
Drehscheibe, Knotenpunkt – Effective centre of an activity, region, or network.

hub-and-spokes system [hʌb ənd spəʊks ˈsɪstəm]
Transportnetzstruktur – In logistics, the hub is the central node to and from which transportation takes place to all the end nodes, whereas spoke stands for the commodity flow from and to the hub.

Human Development Index (HDI)
[ˈhjuːmən dɪˈveləpmənt ˈɪndeks]
Index der menschlichen Entwicklung – Number which helps to compare the degree of human development in countries based on life expectancy, literacy rate, the educational system, and → GDP per capita.

hurricane [ˈhʌrɪkən]
Hurrikan – System of very strong winds circulating around a low-pressure area in Central and North America, combined with heavy rainfall and thunderstorms.

hydroelectric dam [ˌhaɪdrəʊɪˈlektrɪk dæm]
Wasserkraft-Talsperre – Dam containing generators to make electricity.

hydroelectric energy [ˌhaɪdrəʊˈɪlektrɪk ˈenə(r)dʒi]
Energie aus Wasserkraft – Sustainable energy created by flowing water.

hydrogen fuel cell [ˈhaɪdrədʒən ˈfjuːəl sel]
Brennstoffzelle – → Fuel cell.

hydropower [ˈhaɪdrəʊˈpaʊə(r)]
Wasserkraft – Electrical energy gained by running water.

hypocentre [ˈhaɪpə(ʊ)sɛntə]
Hypozentrum – Area inside the → lithosphere where an → earthquake originates. It is located directly below the → epicentre of an earthquake.

I

igneous rock [ˈɪgnɪəs rɒk]
Magmatisches Gestein – Rock that is formed when → magma cools down and turns solid, turning into → plutonic rock and → volcanic rock.

illumination zone [ɪˌl(j)uːmɪˈneɪʃ(ə)n zəʊn]
Beleuchtungszone – Depending on the angle, the sun's rays heat the surface of the → Earth differently, creating areas with a similar temperature that stretch like belts around the Earth.

immigration [ˌɪmɪˈgreɪʃ(ə)n]
Einwanderung – Movement of people into a country.

import [ˈɪmpɔːt]
Einfuhr (Import) – Process of buying goods or services from another country as well as their financial value.

import good [ˈɪmpɔːt gʊd]
Importgut – Product that is brought into one country from another.

industrial agriculture [ɪnˈdʌstriəl ˈægrɪkʌltʃə(r)]
industrielle Landwirtschaft – Form of → farming that refers to the industrialised production of → livestock, → poultry, fish, and crops which includes innovation in agricultural machinery and farming methods, and genetically modified plants and animals.

industrial park [ɪnˈdʌstriəl pɑːk]
Industriegebiet – Area especially for factories on the edge of a town.

industrialisation [ɪnˈdʌstriəlaɪˈzeɪʃ(ə)n]
Industrialisierung – Large-scale introduction of manufacturing and advanced technical enterprises into an area, society, or country.

infant mortality rate [ˈɪnfənt mɔːˈtæləti reɪt]
Säuglingssterblichkeitsrate – Figure, usually given in per mil (‰), showing the number of deaths of children under age 1 within one year in a certain area.

infiltration [ɪnfɪlˈtreɪʃ(ə)n]
Infiltration – The seepage of water into soil or rock.

informal economy [ɪnˈfɔːml ɪˈkɒnəmi]
Schattenwirtschaft – The production of goods or services to generate, mostly casual, employment and incomes outside the supervision of the state.

informal sector [ɪn'fɔ:ml 'sektə(r)]
informeller Sektor – Sector of the economy which includes all jobs which are not recognised as normal income sources, and on which taxes are not paid.

Information Technology, IT [ɪnfə'meɪʃ(ə)n tɛk'nɒlədʒi, ʌɪ ti:]
Informationstechnologie (IT) – Use of electronic equipment, especially computers, for collecting, storing, and sending out information.

infrastructure ['ɪnfrə,strʌktʃə(r)]
Infrastruktur – Facilities which provide the services for communal and individual life; e. g. transport networks, energy and water supply, communication systems, educational institutions, health care facilities.

in-migration [ɪn maɪ'greɪʃn]
regionale Zuwanderung – The movement of people into a new area from elsewhere in the same country.

inner city ['ɪnə(r) 'sɪti]
Innenstadt – Area of a → core city which is the centre of administration, shopping, business and entertainment.

inner core [ɪnə kɔ:]
Innerer Erdkern – Innermost part of the Earth, mainly consisting of the two metals nickel and iron.

inselberg ['ɪns(ə)lbə:g]
Inselberg – A single isolated rock hill or mountain which rises within a plain.

in-situ leaching (ISL) [,ɪn 'sɪtju: li:tʃɪŋ]
In-situ-Laugung, Untertagelaugung – Leaves the ore 'where it is', in the ground, and recovers the minerals from it by dissolving them with chemicals and pumping the mineral-enriched solution to the surface where the minerals can be recovered.

in-situ mining [ɪn 'sɪtju: 'mʌɪnɪŋ]
in-situ Bergbau – Mining process used to recover → minerals or → bitumen through boreholes drilled into a → deposit, without digging.

integration [ɪntɪ'greɪʃ(ə)n]
Eingliederung – The process of bringing together different groups, regions, or countries to improve economic, political, social, and → environmental cooperation.

International Date Line (IDL) [,ɪntə(r)'næʃ(ə)nəl deɪt laɪn]
Internationale Datumsgrenze – Imaginary line roughly along the 180° → meridian. West of this line, the calendar date is one day ahead of the calendar date east of the line.

Intertropical Convergence Zone (ITCZ) [,ɪntə'trɒpɪk(ə)l kən'və:dʒ(ə)ns zəʊn, ʌɪ ti: si: zɛd]
Innertropische Konvergenzzone – The area north and south of the → Equator where the northeast and southeast → trade winds come together.

irrigation [,ɪrɪ'geɪʃ(ə)n]
Bewässerung – Process of providing plants with extra water, especially in dry areas where the amount of natural → precipitation is not high enough.

J

joint venture [,dʒɔɪnt 'ventʃə(r)]
Joint Venture, gemeinsames Tochterunternehmen – Business activity that involves at least two companies doing something together.

just-in-time production [dʒʌst ɪn taɪm prə'dʌkʃn]
Produktion auf Abruf, Just-in-Time-Produktion – Production of the necessary items in necessary quantities at the necessary time.

K

key industry [ki: 'ɪndəstri]
Schlüsselindustrie – Industry of primary importance to an area's or country's economy. Many industrial activities are directly or indirectly dependent on it.

L

labour force ['leɪbə(r) fɔ:s]
Arbeiterschaft – All the people working for a certain company, or all the potential workers in a certain area.

lahar ['lɑ:hɑ:]
Lahar – Special type of mudflow made up of → volcanic ash, rock debris, and water.

land subsidence [land səb'sʌɪd(ə)ns]
Bodenabsenkung – Sinking of the land surface which occurs when too much → groundwater is pumped out causing the overlaying ground to collapse. It can also occur when minerals are extracted by mining.

land use system [land ju:z 'sɪstəm]
Landnutzungssystem – The way in which the use of a certain area of land is managed.

landfill ['lan(d)fɪl]
Landaufschüttung – Site created by the process of building up new land.

landlocked country ['læn(d),lɒkt 'kʌntri]
Binnenstaat – Country that is completely surrounded by land.

landmass [lændmæs]
Landmasse – Large area of land, such as a → continent, that is wholly or mostly surrounded by water.

landscape ['læn(d),skeɪp]
Landschaft – The physical elements of landforms and the living elements, such as → vegetation, in combination with human elements, such as → land use, of a certain area.

landslide ['lan(d)slʌɪd]
Erdrutsch – Movement of a mass of → soil or rock downwards from a mountain or hill, sometimes causing damage.

lava ['lɑ:və]
Lava – Molten rock coming from the inside of the Earth where it is called → magma.

lean production [li:n prə'dʌkʃn]
schlanke Produktion, Lean Production – Form of production in which costs are cut by cutting out all activities that do not add value to the production process as well as by outsourcing production steps to specialised companies or enterprises.

leeward side [ˈliːwə(r)d saɪd]
Leeseite – Side (mainly of a mountain) which faces away from the oncoming wind direction.

Less Economically Developed Country (LEDC) [les ˌiːkəˈnɒmɪkli dɪˈveləpt ˈkʌntri]
Entwicklungsland – Country where → GDP per capita, literacy rate, food supply, and → life expectancy are low, whereas → population growth, infant mortality, and the share of the agricultural sector are high.

life expectancy [ˈlaɪf ɪkspektənsi]
Lebenserwartung – Figure, given in years, describing the average age a new-born can expect to reach in that area or for that group.

light industry [lɪʌtˈɪndəstri]
Leichtindustrie – Industry that makes small things, mostly for end consumers, does not need large, heavy machinery and has little environmental impact.

lignite [ˈlɪgnaɪt]
Braunkohle – Soft, brown coal, with a low energy content, often having a distinct wood-like texture.

line of latitude [laɪn ɒv ˈlætɪˌtjuːd]
Breitengrad – Set of lines of the → global grid which run from east to west and are labelled "North"(N) and "South"(S).

line of longitude [laɪn ɒv ˈlɒndʒɪˌtjuːd]
Längengrad – Set of lines of the → global grid which run from north to south and are labelled "East"(E) and "West"(W) crossing the → North Pole as well as the → South Pole.

liquefied natural gas (LNG) [ˈlɪkwɪfaɪd ˌnætʃrəl ˈgæs]
verflüssigtes Erdgas – → Natural gas compressed at moderate pressure and cooled to remain liquid and then stored and transported in insulated pressure containers.

lithosphere [ˈlɪθəsfɪə(r)]
Lithosphäre – Solid part of the → Earth's crust and solid → upper mantle.

livestock farming [ˈlaɪvstɒk ˈfɑːmɪŋ]
Tierhaltung, Viehwirtschaft – Rearing of farm animals for the production of milk, meat or other products.

living standard [ˈlɪvɪŋ ˈstændəd]
Lebensstandard – Level of comfort that people have which depends on wealth, available necessities, economic and political stability, political and religious freedom, environmental quality, climate, and safety.

Local Agenda 21 [ˈləʊkl əˈdʒendə ˈtwentiwʌn]
Regionale Agenda 21 – The implementation of → Agenda 21 on a local level by local authorities and non-governmental organisations.

location factor [ləʊˈkeɪʃ(ə)n ˈfæktə(r)]
Standortfaktor – Factor which affects an enterprise in the choice of its site.

low mountain range [ləʊˈmaʊntɪn reɪn(d)ʒ]
Mittelgebirge – Unit of several mountains, usually below 1500 m.

lower mantle [ˈləʊə ˈmant(ə)l]
Unterer Erdmantel – Solid layer of the Earth between the → upper mantle and the → outer core.

lowland(s) [ˈləʊləndz]
Tiefland – Flat, low lying → landscape, sometimes hardly above sea level.

M

magma [ˈmægmə]
Magma – Molten rock inside the Earth that is called → lava when it comes to the surface.

magnitude [ˈmagnɪtjuːd]
Erdbebenstärke – Gives information about how much energy is set free during the event of an → earthquake.

malnutrition [ˌmælnjuːˈtrɪʃn]
Fehlernährung – Deficiencies, excesses or imbalances in the consumption of nutrients.

manufacturer [ˌmænjuˈfæktʃərə(r)]
Hersteller, Produzent – Company producing goods in factories.

manufacturing industry [ˌmænjʊˈfæktʃərɪŋ ˈɪndəstri]
verarbeitende Industrie – Industries engaged in the transformation of materials or components into consumer or industrial goods.

manufacturing site [ˌmænjʊˈfæktʃərɪŋ saɪt]
Produktionsstätte – Place or location where goods are produced in factories.

marine energy [məˈriːn ˈenədʒi]
Meeresenergie – All types of energy present in the ocean or along the coast generated by waves, ocean currents, gravity, temperature or salt gradient.

marine pollution [məˈriːn pəˈluːʃn]
Meeres-/Ozeanverschmutzung – Contamination of the oceans mainly by human activities.

maritime climate [ˈmærɪˌtaɪm ˈklaɪmət]
maritimes Klima – → Climate with little temperature differences between cool summers and mild winters.

maritime climate [ˈmærɪˌtaɪm ˈklaɪmət]
Ozeanische Klima – → Climate influenced by the sea; wet with small variations in seasonal temperatures.

market economy [ˈmɑː(r)kəɪt ɪˈkɒnəmi]
Marktwirtschaft – → Free market economy.

market gardening [ˈmɑː(r)kɪt ˈgɑː(r)d(ə)nɪŋ]
Erwerbsgartenbau – Commercial growing of vegetables, salad, flowers, etc., mostly sold in → cities nearby.

mass tourism [mæs ˈtʊərɪz(ə)m]
Massentourismus – Business of providing services for holiday activities on a large scale.

mechanisation [mɛk(ə)nʌɪˈzeɪʃ(ə)n]
Mechanisierung – Providing machinery, especially to replace human or animal labour.

Mediterranean climate [ˌmedɪtəˈreɪniən ˈklaɪmət]
mediterranes Klima – → Climate with warm, dry summers and mild, wet winters.

megacity [ˈmegəsɪti]
Megacity – → Metropolitan area with a total population of more than 10 million people.

meridian [məˈrɪdiən]
Meridian – One half of a → line of longitude. They connect → North Pole and → South Pole.

metamorphic rock [mɛtə'mɔ:fɪk rɒk]
Metamorphes Gestein – Rock formed when → sedimentary or → igneous rock is chemically transformed due to high pressure and/or high temperatures.

methane ['mi:θeɪn]
Methan – A potent → greenhouse gas which is the main component of → natural gas and a component of → crude oil, as well as a product of some biological processes without oxygen.

metropolitan area [ˌmetrə'pɒlɪtən 'eəriə]
Ballungsraum – Large urban settlement consisting of cities surrounded by urban areas with a total population exceeding one million inhabitants.

microcredit ['mʌɪkrəʊ'krɛdɪt]
Mikrokredit – Financial support for poverty-stricken individuals to help them start their own business.

micro-hydro power station ['maɪkrəʊ 'haɪdrəʊ 'paʊə steɪʃn]
Mini-Wasserkraftwerk – Water is diverted from the stream or river and is redirected back into the same watercourse further downstream, driving a → turbine which generates up to 100 kW electricity.

mid-oceanic ridge [mɪd 'əʊʃ(ə)n rɪdʒ]
Mittelozeanischer Rücken – Mountain range at the bottom of the ocean, indicating the border of two → plates with → constructive margins; created by → continental drift and volcanic activity.

migrant worker ['maɪgrənt 'wɜ:(r)kə(r)]
Arbeitsmigrant – Person who leaves an area in order to work somewhere else.

migration [maɪ'greɪʃn]
Wanderung – The movement of people or animals from one place to another.

migration policy [maɪ'greɪʃn 'pɒləsi]
Migrationspolitik – National policy regulating the level of documented → migration for permanent settlement, temporary work or family reunification into and out of the country.

mining industry ['mʌɪnɪŋ 'ɪndəstr]
Bergbau – Industry which deals with the extraction of → natural resources.

mixed forest [mɪkst 'fɒrɪst]
Mischwald – Forest with both → deciduous trees and → coniferous trees.

monoculture ['mɒnəʊˌkʌltʃə(r)]
Monokultur – Cultivation of one main crop, usually on large fields or → plantations.

monsoon [mɒn'su:n]
Monsun – Wind system that changes directions seasonally, creating → wet seasons and → dry seasons, especially in South Asia and coastal areas of West Africa.

moon [mu:n]
Mond – Object that circles the → Earth.

More Economically Developed Country (MEDC)
[mɔ:(r) ˌi:kə'nɒmɪkli dɪ'veləpt 'kʌntri]
wirtschaftlich stärker entwickeltes Land – Country that characteristically shows high → GDP per capita, literacy rate, food supply, → life expectancy and share of the → tertiary sector, as well as low → population growth, low → infant mortality, and a low share of the → primary sector.

mosque [mɒsk]
Moschee – Place of Muslim worship that usually has a minaret.

multinational company [ˌmʌlti'næʃnəl 'kʌmpəni]
multinationale Gesellschaft, Multi – A company operating in several countries, but managed from one (home) country.

N

national economy ['næʃnəl ɪ'kɒnəmi]
Volkswirtschaft – Economy of an entire country, including its financial resources and management and including the value of all goods and services manufactured within it.

national park ['naʃ(ə)n(ə)l pɑ:k]
Nationalpark – An area mainly used for the conservation of undisturbed nature.

natural gas [ˌnætʃrəl 'gæs]
Erdgas – Mixture of gases (→ methane, nitrogen, carbon dioxide, etc.) located near coal and → crude oil deposits.

natural resource ['nætʃrəl rɪ'sɔ:s]
natürliche Ressource – Natural materials or capacity of an area, consisting of water, land, forests, mineral deposits, etc. which have economic value.

net migration gain [net maɪ'greɪʃn geɪn]
Nettozuwanderung – More people immigrating to, than emigrating from, a region.

net migration loss [net maɪ'greɪʃn lɒs]
Nettoabwanderung – More people emigrating from, than immigrating to, a region.

new town [nju: taʊn]
New Town, Neue Stadt – A planned town close to an already existing town or city created in order to relocate → population and businesses.

Newly Industrialised Country (NIC)
['nju:li ɪn'dʌstriəlaɪzd 'kʌntri]
Schwellenland – Country where industry has been growing quickly for the first time; expected to be in a transition between→ LEDCs and → MEDCs.

NGO (Non-Governmental Organisation)
[ɛn dʒi: əʊˌnɒn ˌgʌvə(r)n'ment(ə)l ˌɔ:(r)gənaɪ'zeɪʃ(ə)n]
Nichtregierungsorganisation, NGO – Mostly privately financed, non-profit, voluntary citizens' group, often organised around specific issues, such as human rights, → environment or health.

nomad ['nəʊmæd]
Nomade – Person without a permanent home moving with a herd from one place to another seeking pasture and / or water for the animals.

non-conventional gas [nɒn kən'venʃənl 'gæs]
Gas aus unkonventionellen Lagerstätten – → Natural gas found in unusual underground situations such as very impermeable reservoirs, hydrates, and coal → deposits.

non-renewable energy [nɒn kən'venʃənl 'gæs]
nichterneuerbare Energie – Natural resource such as a → fossil fuel or uranium, which, once consumed, cannot be replaced.

non-renewable resource [nɒn rɪ'nju:əb(ə)l rɪ'zɔ:(r)s]
nicht erneuerbarer Rohstoff – → Natural resource that exists in limited supply and cannot be recovered after its use, mostly → non-renewable energy.

norther ['nɔ:θə]
Norther – Strong cold north wind blowing in autumn and winter over Texas and the Gulf of Mexico.

Northern Hemisphere ['nɔ:(r)ðə(r)n 'hemɪˌsfɪə(r)]
Nordhalbkugel – Half of the Earth north of the → Equator.

nuclear disaster ['nju:kliə(r) dɪ'zɑ:stə(r)]
Atomunfall – Incident that results in the release of radioactive material from a nuclear facility or container.

nuclear energy ['nju:kliə(r) 'enədʒi]
Kernenergie – Energy that is released by splitting apart or fusing the nuclei of atoms.

nuclear fuel cycle ['nju:kliə(r) 'fju:əl 'saɪkl]
Kernbrennstoffkreislauf – Radioactive uranium and plutonium, after being used to produce electricity, need storing or reprocessing and recycling to be used again in another cycle.

nuclear power station ['nju:kliə(r) 'paʊə steɪʃn]
Kernkraftwerk – Factory where → nuclear energy is converted into heat and electricity.

nutrient cycle ['nju:triənt 'saɪk(ə)l]
Nährstoffkreislauf – Circulation of → minerals taken up by the roots of plants, used by plants, animals, and humans and returned into the soil after the decay of plants by bacteria and fungi.

O

ocean ['əʊʃ(ə)n]
Ozean – Large water mass between the → continents.

oceanic crust [ˌəʊʃi'ænɪk krʌst]
ozeanische Kruste – Thinner but heavier part of the → Earth's crust.

offshore [ˌɒf'ʃɔ:(r)]
auf offener See – Away from the coast, at sea.

offshore oil platform (oil rig)
[ˌɒf'ʃɔ:(r) ɔɪl 'plætˌfɔ:(r)m (ɔɪl rɪg)]
Ölbohrinsel – Large structure in the sea with facilities to drill wells, to extract and process oil and → natural gas, and to temporarily store them.

oil crisis [ɔɪl 'kraɪsɪs]
Ölkrise – Sudden drop in the supply of oil that is accompanied by a rise in price.

oil field [ɔɪl fi:ld]
Ölfeld – Deposits of → crude oil in the → Earth's crust.

oil industry [ɔɪl 'ɪndəstri]
Ölindustrie – Deals with the exploration, extraction, refining, transporting as well as marketing of oil and oil products.

oil sand [ɔɪl sand]
Ölsand – Sand and rock material which contains crude → bitumen.

oil spill [ɔɪl spɪl]
Ölverschmutzung – Accident in which oil has been lost during transport and caused pollution.

oil terminal [ɔɪl 'tɜ:(r)mɪn(ə)l]
Ölterminal – Large industrial complex where oil and → natural gas are stored and from which they are transported elsewhere.

oil well ['ɔɪl wel]
Ölquelle – A hole drilled into the ground for the extraction of petroleum.

oilseed rape [ˌɔɪlsi:d 'reɪp]
Raps – Bright yellow flowering → crop, cultivated mainly for its oil-rich seed.

old industrial region [əʊld ɪn'dʌstriəl 'ri:dʒən]
Altindustrieregion – Industrial region which was often based on mining and heavy industry. Most of these regions have undergone some form of rapid decline in the second half of the 20th century and may be in the process of structural change.

onshore ['ɒnʃɔ:(r)]
an Land – On land.

OPEC ['əʊpek]
OPEC – Organisation of oil exporting countries with 14 member states (2017).

open-pit mining, open-cast mining
['əʊpən pɪt 'maɪnɪŋ', əʊpən kɑ:st 'maɪnɪŋ']
Tagebau – Minerals are produced through big holes near the surface, not from deep under the ground.

organic farming [ɔ:'gænɪk 'fɑ:mɪŋ]
biologische Landwirtschaft – Sustainable form of agriculture relying on natural inputs into farming while refraining from artificial ones like fertilizer and pesticides.

outer core ['aʊtə kɔ:(r)]
Äußerer Erdkern – Liquid layer of the Earth between the solid → lower mantle and the solid → inner core.

out-migration [aʊt maɪ'greɪʃn]
Abwanderung – Movement of people away from one place in a country towards another one in the same country.

outskirts ['aʊtˌskɜ:(r)ts]
Stadtrand, Außenbezirk – Outlying areas of a → city or town.

outsourcing ['aʊtsɔ:sɪŋ]
Outsourcing, Auslagerung – Transfer of tasks, services or the manufacturing of products to a company from outside typically used to save costs.

overfishing [ˌəʊvə'fɪʃɪŋ]
Überfischen – Fishing at a rate that exceeds the rate which is necessary to replenish the → fish stock.

overgrazing ['əʊvə(r)greɪzɪŋ]
Überweidung – Depletion of an areas → vegetation by grazing too big herds which can result in erosion of the grassland and → desertification.

overnutrition ['əʊvənju'trɪʃn]
Überernährung – Consuming too much food and falling ill from it.

overpopulation [əʊvə'pɒpjʊleɪt]
Überbevölkerung – Too much → population in an area causes → environmental problems and an impaired → quality of life.

P

pack ice [pak ʌɪs]
Packeis – Mass of ice floating in the sea, formed by smaller pieces of ice freezing together.

package tour ['pækɪdʒ tʊə(r)]
Pauschalreise – Tour arranged by a travel agent. Transportation, food, and lodging are all provided at an inclusive price.

parent rock ['pɛ:r(ə)nt rɒk]
Ausgangsgestein – Rock from which a soil is formed by outer influences (e.g. → weathering).

parliament ['pɑ:(r)ləmənt]
Parlament – Elected group of people who pass laws, control the government, and discuss the political affairs of a country (e.g. the German Bundestag).

pasture ['pɑ:stʃə]
Weide – Farmland that consists mainly of grasses and is grazed by domesticated → livestock.

permafrost ['pə:məfrɒst]
Permafrost – Perennially frozen soil in arctic and subarctic regions.

pesticide ['pɛstɪsʌɪd]
Pestizid, Pflanzenschutzmittel – Substance to kill pests, i.e. harmful insects, rodents, or weeds.

petrochemical industry [ˌpetrəʊ'kemɪkl 'ɪndəstri]
Petrochemische Industrie – Heavy industry producing synthetic materials, mainly from → refinery products and → natural gases.

photovoltaic (PV) ['fəʊtəʊvɒltɑ:ɪk]
Photovoltaik – Direct conversion of solar energy into electrical energy through → solar cells.

pipeline ['paɪplaɪn]
Pipeline, Rohrleitung – Long tube with pumps, valves, and control devices to transport liquids, gases, or finely dispersed solids.

place of interest [pleɪs ɒv 'ɪntrəst]
Sehenswürdigkeit – Place famous for its scenery or famous historical site.

plain [pleɪn]
Ebene, Flachland – Area of flat land.

planet [plænet]
Planet – Celestial body that goes around the → Sun.

planned economy [planned ɪ'kɒnəmi]
Planwirtschaft – Economic system with a government interfering in business activities.

plant [plɑ:nt]
hier: Produktionsanlage, Fabrik – Building or group of buildings for the manufacture of a product; a factory.

plantation [plɑ:n'teɪʃ(ə)n]
Plantage – Farm in the → Tropical Zone where → cash crops are produced on a large scale which usually specialise on a single → crop.

plastic ['plæstɪk]
Plastik, Kunststoff – Material consisting of synthetic organic compounds, most commonly derived from petrochemicals, but also from renewable materials.

plate [pleɪt]
Platte (der Lithosphäre) – Earth's → lithosphere is broken into pieces which consist of → oceanic and → continental crust.

plate tectonics [pleɪt tɛk'tɒnɪks]
Plattentektonik – Widely accepted theory that the → lithosphere of the → Earth is broken into → plates which move on the viscous → upper mantle.

plateau ['plætəʊ]
Plateau – Rather flat area which is higher above sea level than the surrounding land.

plutonic rock [plu:'tɒnɪk rɒk]
Tiefengestein – → Igneous rock that formed inside of the Earth.

points of the compass [pɔɪnts əv ði: 'kʌmpəs]
Himmelsrichtungen – Mark the directions north, south, east, west.

point-to-point system [pɔɪnt tə pɔɪnt 'sɪstəm]
Punkt-zu-Punkt System – Transport system (air, rail, road, water) linking destinations directly.

polar day ['pəʊlə(r) deɪ]
Polartag – Period lasting between one day at the polar circles and six months at the poles, in which north of the → Arctic Circle (or south of the → Antarctic Circle), the Sun stays above the horizon, so that it does not get dark.

polar night ['pəʊlə(r) nʌɪt]
Polarnacht – Period lasting between one day at the polar circles and six months at the poles, in which north of the → Arctic Circle (or south of the → Antarctic Circle), the Sun stays below the horizon, so that it does not get light.

Polar Zone ['pəʊlə(r) zəʊn]
Polarzone, Polargebiet – Zone between the → Arctic Circle or → Antarctic Circle and their respective poles.

pollution [pə'lu:ʃn]
Verschmutzung – Contamination of the → environment or parts of it (soil, air, water) caused by human activity.

population [ˌpɒpjʊ'leɪʃ(ə)n]
Bevölkerung – Total number of people living in a specific area.

population density [ˌpɒpjʊ'leɪʃ(ə)n 'densəti]
Bevölkerungsdichte – Total number of people per square kilometre or per square mile in a given administrative area.

population distribution [ˌpɒpju'leɪʃn ˌdɪstrɪ'bju:ʃn]
Bevölkerungsverteilung – Allocation of people in a certain area.

population growth [ˌpɒpjʊ'leɪʃ(ə)n grəʊθ]
Bevölkerungswachstum – Increase or decrease of the → population in a given area.

population growth rate [ˌpɒpju'leɪʃn grəʊθ reɪt]
Bevölkerungswachstumsrate – Figure, given in per cent (%), describing the increase or decrease in the → population of a given area.

population structure [ˌpɒpju'leɪʃn 'strʌktʃə(r)]
Bevölkerungsstruktur – Age and sex composition of a → population or population group in a country, region or city.

poultry farming ['pəʊltri 'fɑ:mɪŋ]
Geflügelzucht, -haltung – Raising of birds commercially, primarily for meat and eggs.

poverty line ['pɒvə(r)ti laɪn]
Armutsgrenze – Minimum level of income necessary for a person to achieve an adequate standard of living. If a person's income is below this level, he or she is said to live in poverty.

power station ['paʊə steɪʃn]
Kraftwerk – Factory where electricity is generated.

power-to-gas (P2G) ['paʊə tə 'gæs]
Power-to-Gas – Technologies that use surplus electrical power, mostly from renewable resources, to make hydrogen from water or methane from water and CO_2.

precipitation [prɪˌsɪpɪ'teɪʃ(ə)n]
Niederschlag – Condensed water vapour from the → atmosphere which reaches the surface of the earth as water or ice.

precision farming [prɪ'sɪʒn 'fɑ:mɪŋ]
Precision Farming, Präzisionslandwirtschaft – Farm management that uses specialised equipment, software and IT services accessing real-time data about the conditions on the field to ensure profitability, sustainability, and protection of the environment.

primary energy ['praɪməri 'enədʒi]
Primärenergie – Any → renewable or → non-renewable energy that has not been subjected to a conversion or transformation process.

primary sector ['praɪməri 'sektə(r)]
primärer Sektor – Sector of an economy making direct use or exploiting → natural resources consisting of agriculture, forestry, fishing and mining.

Prime Meridian [praɪm mə'rɪdiən]
Nullmeridian – Line of 0° longitude which runs through Greenwich, London from where the other longitudes from 0° to 180° east, or from 0° to 180° west, are counted.

public transport ['pʌblɪk 'trænspɔ:(r)t]
Öffentlicher Verkehr – System of buses, trains, etc., running on fixed routes, on which the public may travel.

pull factor [pʊl 'fæktə(r)]
Pull-Faktor – Social, political, economic, or environmental conditions which attract people to a new location.

pumped-storage power station
[pʌmpt 'stɔ:rɪdʒ 'paʊə steɪʃn]
Pumpwasserkraftwerk – → Power station using power at a period of low electricity demand to pump water up to a high storage reservoir so that it can be released to generate electricity at a period of high demand.

push factor [pʊʃ 'fæktə(r)]
Push-Faktor – Social, political, economic, or environmental conditions which make people leave their old location.

Q

quality of life ['kwɒləti əv 'laɪf]
Lebensqualität – Highly subjective measure of health, comfort, and happiness that often include financial security, job satisfaction, family life, and personal safety.

quaternary industry ['kwɒtərˌnɛri 'ɪndəstri]
Quartärer Wirtschaftszweig – Economic activities mainly concerned with education, research and the media, often still regarded as part of the → tertiary sector.

R

radioactive waste [ˌreɪdiəʊ'æktɪv weɪst]
Atommüll – Harmful radiation emitting by-products from the operation of a nuclear reactor or from depleted radioactive material elsewhere.

ragpicker [ræg'pɪkə(r)]
Lumpensammler – Person who makes a living by collecting rags and other waste material from the streets, dumps, etc.

rain gauge [reɪn geɪdʒ]
Niederschlagsmesser – Measures the amount of → precipitation over a certain period of time.

rain shadow ['reɪn 'ʃadəʊ]
Regenschatten – Area on the → leeward side of a mountain, where there is less rainfall due to warming, descending air masses.

rainforest ['reɪnˌfɒrɪst]
Regenwald – Dense forest, often close to the → Equator as → tropical rainforest, in which the climate is hot and wet.

rainy season ['reɪni 'si:z(ə)n]
Regenzeit – Time of the year when most of a region's average annual → precipitation occurs.

rate of natural increase [reɪt əv 'nætʃrəl ɪn'kri:s]
natürliche Wachstumsrate – → Birth rate minus → death rate.

raw material [rɔ: mə'tɪəriəl]
Rohstoff – Unprocessed material used for the production of goods.

reclamation of land [ˌreklə'meɪʃ(ə)n əv lænd]
Landgewinnung, Urbarmachung von Land – Process of creating new land from ocean, riverbeds, or lakes.

recycling [ˌri:'saɪklɪŋ]
Recycling – Process of treating things that have already been used so that they can be used again.

refinery [rɪ'faɪnəri]
Raffinerie – Industrial plant for purifying a crude substance, e.g. petroleum or sugar.

reforestation [ˌri:fɒrɪ'steɪʃ(ə)n]
(Wieder-)Aufforstung – Planting of trees in areas where the original forest has been cleared. It is used to aid the → environment, e.g. as an important method to reduce the danger of → soil erosion.

regional disparity ['ri:dʒənl dɪ'spærəti]
regionale Ungleichheit – Difference between areas or regions in terms of population density, economic power, natural conditions, etc.

regional policy ['ri:dʒənl 'pɒləsi]
Regionalpolitik – Policy concerned with promoting a balanced distribution of industrial activity, employment and wealth between the various regions of a country.

reindustrialisation [riːɪnˌdʌstriəlaɪˈzeɪʃn]
Reindustrialisierung – Renewed introduction of industry to a region which has lost its industrial capacity.

relief [rɪˈliːf]
Relief – Shape of the Earth's surface.

relief rainfall [rɪˈliːf ˈreɪnˌfɔːl]
Reliefregen, Steigungsregen – Rainfall caused by rising air masses which cool down crossing mountain ranges. This leads to the condensation of water vapour and → precipitation.

renewable energy, renewable resource [nˈnjuiabl ˈenədʒi, nˈnjuiabl rɪˈsɔːs]
erneuerbare Energie – Energy from → resources which can replenish, such as sunlight, wind, rain, tides, waves, and geothermal heat.

reprocessing facility [riːˈprəʊsesɪŋ fəˈsɪləti]
Wiederaufbereitungsanlage – Factory to separate and recover energy-rich plutonium and uranium from spent nuclear → fuel.

research and development (R&D)
[rɪˈsɜː(r)tʃ ænd dɪˈveləpmənt, ɑː ənd diː]
Forschung und Entwicklung – Term covering basic research, applied research, and experimental development.

reserve [rɪˈzɜːv]
Reserve – Part of a → resource that can be recovered economically with the available technologies.

reservoir [ˈrezə(r)ˌvwɑː(r)]
Stausee – Artificial lake created by building a → dam to store water for → irrigation, → fresh water supply, producing power, recreational purposes, etc.

residential quarter, residential area
[ˌrezɪdenʃ(ə)l ˈk(w)ɔːtə, ˌrezɪdenʃ(ə)l ˈeəriə]
Wohngebiet – Area which is used for building private houses.

resort [rɪˈzɔː(r)t]
Urlaubsort – Place with facilities for people on holiday.

resource [rɪˈsɔːs]
Ressource – Total amount of undiscovered and discovered material that exists.

retirement age [rɪˈtaɪəmənt eɪdʒ]
Rentenalter – Age at which most people normally stop working.

revitalisation [ˌriːˌvaɪtəlaɪˈzeɪʃn]
Revitalisierung, Wiederbelebung – Process of making a region or economy stronger, more active or more productive again.

river bank [ˈrɪvə(r) baŋk]
Flussufer – Area of land alongside a river.

river mouth [ˈrɪvə(r) maʊθ]
Flussmündung – Site where a river flows into an ocean, a sea, a lake, or another river.

rock cycle [rɒk ˈsʌɪk(ə)l]
Gesteinskreislauf – Cycle that describes the formation and transition of different types of rock from others.

run-of-the-river power station
[rʌn əv ðə ˈrɪvə(r) ˈpaʊə steɪʃn]
Laufwasserkraftwerk – → Power station utilising the river flows to generate power.

rural area [ˈrʊərəl ˈeəriə]
ländliches Gebiet, "Land" – Countryside; in contrast to an → urban area.

rural-to-urban migration [ˈrʊərəl tə ˈɜːbən maɪˈgreɪʃn]
Land-Stadt-Wanderung – The movement of people from the countryside to city areas.

S

salinisation [səˈlɪnaɪ zeɪʃ(ə)n]
Versalzung – Process of making a soil infertile by increasing its salt content.

savanna [səˈvænə]
Savanne – Grassland in the → Tropical Zone, between the → desert and the → tropical rainforest, with distinct dry and rainy seasons. There are three types: wet savanna, dry savanna, and bush savanna.

Schengen Agreement [ˈʃeŋən əgriːmənt]
Schengener Abkommen – Treaty signed in 1985 which created the so-called 'Schengen Area' which is an area of 26 European states in which border checks have officially been abolished.

seafloor spreading [siːflɔː sprɛdɪŋ]
Meeresbodenspreizung – Process in which new → oceanic crust is formed along → constructive margins, typically found in the middle of oceans and accompanied by volcanic activity and → mid-oceanic ridges.

season [ˈsiːz(ə)n]
Jahreszeit – Period within a year showing specific natural conditions which are caused by the duration of daylight (spring, summer, autumn, winter), and/or the amount of rainfall (→ wet season, → dry season).

secondary energy [ˈsekəndri ˈenədʒi]
Sekundärenergie – Energy which is transformed from primary energies, like electricity, gasoline, or methane.

secondary sector [ˈsekəndri ˈsektə(r)]
sekundärer Sektor – Production involving the conversion of raw materials into finished or semi-finished goods which includes all manufacturing, processing and construction.

sediment [ˌsedimenˈt]
Sediment – Pieces of weathered and eroded rock that are transported and deposited.

sedimentary rock [ˌsedimenˈtəri rɒk]
Sedimentgestein – Solidified → sediment.

sedimentation [ˌsedimenˈteɪʃ(ə)n]
Sedimentation – Deposition of material (mud, sand) in rivers, lakes, depressions, etc.

segregation [ˌsegrɪˈgeɪʃn]
Segregation, Trennung – Distribution of specific groups of people into distinct areas, usually based on, e.g. race, religion, or income.

serir ['sɪəri(r)]
Geröll-, Kieswüste – → Desert landscape covered with gravel and pebbles, which makes up about 10 per cent of the Sahara Desert.

service industry ['sɜ:(r)vɪs 'ɪndəstri]
Dienstleistungsgewerbe – Economic activity concerned with the offer of any services, e. g. trade, transport; → tertiary industry.

sewage ['su:ɪdʒ]
Abwasser – Water that has been affected in quality mostly by human influences.

shield volcano [ʃi:ld vɒl'keɪnəʊ]
Schildvulkan – Special type of → volcano which is created by the flow of runny → lava from a single → vent.

shifting cultivation ['ʃɪftɪŋ ˌkʌltɪ'veɪʃ(ə)n]
Wanderfeldbau – Pattern of agricultural land use mainly in the → Tropical Zone, which is characterised by the → cultivation of new farmland at another place when yields decline at the old location.

shopping centre ['ʃɒpɪŋ sentə(r)]
Einkaufszentrum – Form of shopping in which one or more buildings form a complex of shops with interconnecting passages that enable customers to walk everywhere inside.

skilled worker [skɪld 'wə:kə]
Facharbeiter – Worker who has acquired special skills or who works at a special occupation.

slash-and-burn [slæʃ ænd bɜ:(r)n]
Brandrodung – Method of → farming, e. g. in the → tropical rainforest where farmland is reclaimed by burning down the original → vegetation.

slum [slʌm]
Elendsviertel, Slum – Area with primitively built or rundown houses where very poor people live.

smallholding ['smɔ:lhəʊldɪŋ]
Kleinbäuerlicher Betrieb – Farms supporting a single family with a mixture of → cash crops and → subsistence farming.

smart grid [smɑ:t grɪd]
intelligentes Stromnetz – An electricity network based on digital technology that supplies electricity to consumers via two-way automated communication to help improve efficiency, reduce energy consumption and costs.

smog [smɒg]
Smog – Fog made heavier and darker by smoke and chemical fumes. Usually found in cities.

soil [sɔɪl]
Boden – Uppermost layer of the → Earth's surface made up of a mixture of weathered rock material, decomposed organic matter, microorganisms, air and water.

soil compaction [sɔɪl kəm'pæktʃn]
Bodenverdichtung – Increase of density, or decrease in porosity, of soil.

soil conservation [sɔɪl ˌkɒnsə'veɪʃn]
Bodenschutz – Steps to stop and prevent → soil degradation from erosion and reduced fertility.

soil degradation [sɔɪl ˌdegrə'deɪʃ(ə)n]
Bodendegradierung – Process of declining → soil fertility caused by erosion, often accelerated by human activities.

soil erosion [sɔɪl ɪ'rəʊʒ(ə)n]
Bodenerosion – Naturally occurring process of the displacement of upper layer of soil by water and wind which might be accelerated by human activities.

soil fertility [sɔɪl fə:'tɪlɪti]
Bodenfruchtbarkeit – Ability of a soil to supply plant growth.

solar cell [ˌsəʊlə 'sel]
Solarzelle – Small appliance designed to convert sunlight into electrical energy, mostly consisting of layers of specially prepared silicon.

solar collector [ˌsəʊlə kə'lektə(r)]
Sonnenkollektor – Device to capture and use solar radiation for heating air or water.

solar energy [ˌsəʊlə 'enədʒi]
Sonnenenergie – Energy the → Earth receives from the sun's rays.

solar panel [ˌsəʊlə 'pænl]
Solarmodul – Electrical implement consisting of a large array of connected → solar cells.

solar system ['səʊlə(r) 'sɪstəm]
Sonnensystem – Sum of a sun, the surrounding → planets, and their → moons.

solar tower [ˌsəʊlə 'taʊə(r)]
Solarturm – Tower in which light is gathered from surrounding mirrors to produce → concentrated solar power.

source [sɔ:s]
Quelle (hier: Wasserquelle) – Site where an impermeable layer comes close to the surface (normally at a slope), so that → groundwater escapes forming the beginning of a river.

South Pole [saʊθ pəʊl]
Südpol – The southernmost point of the → Earth located at 90° S.

Southern Hemisphere ['sʌðə(r)n 'hemɪˌsfɪə(r)]
Südhalbkugel – Half of the → Earth south of the → Equator.

special economic zone (SEZ) ['speʃl ˌi:kə'nɒmɪk zəʊn]
Sonderwirtschaftszone – Area with particular economic advantages, for example lower taxes compared with the rest of the country, to encourage investment and development there.

specialisation [ˌspeʃəlaɪ'zeɪʃn]
Spezialisierung – Method of production where a business, area, or → economy focusses production on a limited variety of products or services to gain greater efficiency.

standard of living ['stændəd əv 'lɪvɪŋ]
Lebensstandard – Level of material comfort as measured by the goods, services, and luxuries available to an individual, group, or nation.

state [steɪt]
Staat – An independent political unit, e. g. a country or a political unit within a country.

steel crisis [sti:l ˈkraɪsɪs]
Stahlkrise – Crisis of the steel industry between the 1970s and the 1990s which led to the decline of the → old industrial regions of Europe and the USA due to falling steel prices.

steppe [stɛp]
Steppe – Grassland in the → Temperate Zone and → Subtropical Zone with dry periods in summer and an annual average → precipitation below 600 mm.

stratovolcano [ˌstratəʊvɒlˈkeɪnəʊ]
Stratovulkan – Type of → volcano which is created by the deposition of erupted ashes and → lava on top of each other, forming a cone.

structural change [ˈstɒk ɪkstʃeɪndʒ]
Strukturwandel – Change in the socio-economic structure of an area over a longer period of time, in MEDCs after the decline of a major industry and the introduction of other types of industry.

structural funds[ˈstrʌktʃərəl fʌnds]
Strukturfonds – → EU's development programme which is designed to narrow the gap between the rich and poor parts of the EU.

subduction zone [səbˈdukʃ(ə)n zəʊn]
Subduktionszone – Site where → oceanic crust slides under → continental crust, often the → hypocentre of → earthquakes.

Subpolar Zone [ˈsʌbˈpəʊlə zəʊn]
Subpolare Zone – Transition zone between the → Polar and the → Temperate Zone.

subsidy [ˈsʌbsədi]
Subvention, Fördermittel – Benefit for groups, companies or individuals granted by the government, usually in form of a grant or a reduction of taxes.

subsistence farming [səbˈsɪstəns ˈfa:(r)mɪŋ]
Selbstversorgung durch Landwirtschaft – Growing crops not for profit but mainly to provide the farmer and his family with food.

Subtropical Zone [ˈsʌbˈtrɒpɪk(ə)l zəʊn]
Subtropische Zone – A transition zone between the → Tropical and the → Temperate Zone.

suburb [ˈsʌbɜ:(r)b]
Vorort – District situated on the → outskirts of a → city or town.

suburbanisation [sʌbˌɜ:bənaɪˈzeɪʃn]
Suburbanisierung – Movement of population and economic activities from within cities to the → outskirts.

suburbia [səˈbə:bɪə]
Stadtumland – All suburbs and their inhabitants together.

supply industry [səˈplaɪ ˈɪndəstri]
Zulieferindustrie – Industries which provide other industries with goods, e. g. components which they need.

sustainability [səˌsteɪnəˈbɪləti]
Nachhaltigkeit – Balance between economic, social and ecological activities with the aim to secure → resources for future generations.

sustainable development [səˈsteɪnəbl dɪˈveləpmənt]
nachhaltige Entwicklung – Improving our → quality of life without damaging the quality of life of future generations which involves all individuals, and careful planning at local, regional and global levels.

Sustainable Development Goals (SDGs)
[səˈsteɪnəbl dɪˈveləpmənt goʊls]
Ziele für nachhaltige Entwicklung – Political targets established by the United Nations in order to secure worldwide → sustainable development economically, socially and ecologically.

sustainable forestry [səˈsteɪnəbl ˈfɒrɪstri]
nachhaltige Forstwirtschaft – Environmentally appropriate, socially beneficial, and economically possible management of forests for present and future generations.

swamp [swɒmp]
Sumpf, Moor – Area of wet, spongy land, not usable for → cultivation.

sweatshop [ˈswetˌʃɒp]
Ausbeuterbetrieb – Small business, factory, etc. with poor working conditions, e. g. hard work and low wages.

T

taiga [ˈtʌɪgə]
Taiga – Subpolar → vegetation zone with mainly → coniferous trees south of the tundra in the → Northern Hemisphere.

tanker [ˈtæŋkə(r)]
Tankschiff – Ship designed for the transport of liquids or gases.

taxation [tækˈseɪʃn]
Besteuerung – Charge imposed by the government on the citizens and companies of a country in order to finance the government's expenses.

Temperate Zone [ˈtemp(ə)rət zəʊn]
Gemäßigte Zone – Zone between the → Subtropical Zone and the → Polar Zone, extending between about 40° latitude and the → Arctic or → Antarctic Circle.

temperature inversion [ˈtɛmp(ə)rətʃə ɪnˈvə:ʃ(ə)n]
Temperaturumkehr – Increase of temperature with height whereas usually temperature decreases with height.

temperature range [ˈtemprɪtʃə(r) reɪndʒ]
Temperaturbereich – The difference between the high and low temperature measured over a certain period of time (day, year).

terms of trade [tɜ:mz əv treɪd]
Terms of Trade – Measurement of how many units of imports a country can get for a unit of exported goods.

terrace farming [ˈtɛrəs ˈfa:mɪŋ]
Terrassenfeldbau – Level fields created on the slope of mountains which maximises the → arable land area and reduces → soil erosion and water loss.

tertiary sector [ˈtɜ:ʃəri ˈsektə(r)]
Tertiärer Sektor – Sector of the economy concerned with the selling of services and skills; → service sector, sometimes subdivided in tertiary and → quaternary sector.

tidal barrage [ˈtaɪdl ˈbæra:ʒ]
Gezeitensperrmauer – A dam-like structure used to capture the energy from water moving in and out of a bay or river due to tidal forces.

tidal energy ['taɪdl 'enədʒi]
Gezeitenenergie – → Sustainable energy obtained by harnessing energy from the shifting water in ocean tides.

tilt [tɪlt]
Neigung (der Erdachse) – The → Earth's axis is not vertical. It is oblique with an angle of 23.5° against the Sun.

time difference [taɪm 'dɪfrəns]
Zeitunterschied, Zeitverschiebung – Number of hours between places in two → time zones.

time zone [taɪm zəʊn]
Zeitzone – Division of the → Earth, usually in 15° longitude sections creating 24 such divisions within which the mean time of the central → meridian represents the time for the whole zone.

topsoil [tɒpsɔɪl]
Oberboden – Upper layer of soil (5 to 20 cm) which has the highest concentration of organic matter and microorganisms.

tornado [tɔ:'neɪdəʊ]
Tornado – Very destructive, short-lived, revolving storm, usually together with rain and thunder, mainly in continental USA.

tourist industry, tourism ['tʊərɪst 'ɪndəstri, 'tʊərɪz(ə)m]
Touristikbranche – People, activities, and organisations involved in providing services for people on holiday.

trade [treɪd]
Handel – Activity of buying, selling, or exchanging goods.

trade wind [treɪd wɪnd]
Passatwind – Constant easterly wind which blows from the subtropical high pressure area to the low pressure areas near the → Equator.

transitional climate [træn'zɪʃ(ə)nəl 'klaɪmət]
Übergangsklima – → Climate between → maritime and → continental climate in Europe.

Treaty of Maastricht ['tri:ti əv ma:strɪtʃt]
Maastrichter Vertrag – Charter signed in 1992 which founded the → EU becoming effective on November 1st, 1993 and laying the foundation for a union on an economic, political, judiciary and military level.

tributary ['trɪbjʊt(ə)ri]
Zufluss, Nebenfluss – River or stream which flows into a larger one.

Tropic of Cancer ['trɒpɪk əv 'kænsə(r)]
nördlicher Wendekreis – Line of latitude at 23.5°; north of this line the Sun can never be seen in the → zenith.

Tropic of Capricorn ['trɒpɪk əv 'kæprɪˌkɔ:(r)n]
südlicher Wendekreis – Line of latitude at 23.5° S; south of this line the Sun can never be seen in the → zenith.

tropical rainforest ['trɒpɪk(ə)l 'reɪnˌfɒrɪst]
Tropischer Regenwald – Evergreen → vegetation zone, which is a distinctly layered → rainforest close to the → Equator.

Tropical Zone ['trɒpɪk(ə)l zəʊn]
Tropen – Zone between the → Tropic of Cancer and the → Tropic of Capricorn in which the sun is in the → zenith two times a year.

tsunami [tsu:'nɑ:mi]
Tsunami – Very large and destructive ocean wave, usually caused by an → earthquake under the ocean floor.

tundra ['tʌndrə]
Tundra – Polar and subpolar → vegetation zone where short and cool summers and the → permafrost soil do not allow the growth of trees but instead of grass, moss, lichen, and small bushes.

U

unconventional oil [ˌʌnkən'venʃənl ɔɪl]
Öl aus unkonventionellen Ressourcen – Crude oil that is not produced by traditional extraction methods including offshore oil, oil sands, and tight oil.

undernutrition ['ʌndənju'trɪʃn]
Unterernährung – Lack of food or a lack of the right type of food over a period of time which can lead to poor health, a limited capacity to work and/or death by starvation.

unskilled worker [ʌn'skɪld 'wə:kə]
ungelernte Arbeitskraft – Someone with no special training or experience for performing his or her work adequately.

upland ['ʌplənd]
Hochland – High lying areas of the Earth, e. g. → plains and → basins in mountainous regions or hilly and mountainous area with mountain heights under 1500 m.

upper mantle ['ʌpə 'mant(ə)l]
Oberer Erdmantel – Layer of the Earth between the → Earth's crust and the → lower mantle; part of the → Earth's mantle.

urban area ['ɜ:(r)bən 'eəriə]
städtisches Gebiet, "Stadt" – Built-up area related to a → city or a densely populated area; in contrast to a → rural area.

urban sprawl ['ɜ:(r)bən sprɔ:l]
Zersiedlung – Unrestricted or unregulated spread of → urban area.

urbanisation [ˌɜ:(r)bənaɪ'zeɪʃ(ə)n]
Urbanisierung, Verstädterung – Expansion of towns in number, area and → population.

V

vegetation [ˌvedʒə'teɪʃ(ə)n]
Vegetation – → Earth's plant life.

vegetation [ˌvedʒə'teɪʃ(ə)n]
Vegetation – Variety of plants, which depends on the given → climate and → soil.

vegetation period [ˌvedʒə'teɪʃ(ə)n'pɪərɪəd]
Vegetationsperiode /-zeit – Time of the year when → climatic conditions are favourable for plant growth, common to a place or an area.

vegetation zone [ˌvedʒəˈteɪʃ(ə)n zəʊn]
Vegetationszone – Region showing a specific variety of plants, which depends on the given → climate and → soil.

vent [vent]
Schlot – Opening in a → volcano, which leads from the → magma chamber to the surface and through which → lava and hot gases can escape.

virtual power plant [ˈvɜːtʃuəl ˈpaʊə plɑːnt]
virtuelles Kraftwerk – Cluster of controllable, interconnected, and decentralised residential electricity and heat-producing devices using different energy sources (wind, sun, geothermal and ambient heat etc.), installed in residential houses, small enterprises and public facilities etc., for the exchange of heating, cooling, and electricity.

virtual water [ˈvɜːtʃuəl ˈwɔːtə(r)]
virtuelles Wasser – Volume of fresh water used to create the goods and services that we consume and use, measured at the place where they were actually produced or provided.

volcanic eruption [vɒlˈkænɪk ɪˈrʌpʃ(ə)n]
Vulkanausbruch – Process in which hot volcanic material (→ lava, gases, volcanic ash, etc.) is ejected.

volcanic rock [vɒlˈkænɪk rɒk]
Ergussgestein – → Igneous rock that is formed after reaching the surface of the Earth.

volcano [vɒlˈkeɪnəʊ]
Vulkan – Opening in the → Earth's crust, usually forming a peak, through which → lava, volcanic ash, gases, steam, and rocks come to the surface.

W

water availability [ˈwɔːtə(r) əˌveɪləˈbɪləti]
Wasserverfügbarkeit – Hydrologic capacity of a surface water body, groundwater, and other water sources of a certain area to sustain additional water demands.

water cycle [ˈwɔːtə(r) ˈsaɪk(ə)l]
Wasserkreislauf – Circulation of water between Earth and → atmosphere, caused by → evaporation, → condensation and → precipitation.

wave energy [weɪv ˈenə(r)dʒi]
Wellenenergie – Sustainable energy from ocean surface waves, and the capture of that energy to do useful work.

weather [ˈweðə(r)]
Wetter – Interaction of the atmospheric elements → precipitation, → temperature, humidity, atmospheric pressure, cloud cover, and wind over a short period of time and at any place.

weather elements [ˈweðə(r) ˈelɪmənts]
Wetterelemente – Atmospheric elements which contribute to the → weather.

weather station [ˈweðə(r) ˈsteɪʃ(ə)n]
Wetterstation – Place where the weather elements → precipitation, → temperature, wind speed, wind direction, cloud cover, sunshine duration, and air pressure are studied and recorded.

weathering [ˈweðə(r)ɪŋ]
Verwitterung – Breaking down of rocks, → soil and → minerals by physical and/or chemical processes.

wind energy [ˈwɪnd ˈenədʒi]
Windkraft – Energy contained in blowing wind.

wind farm [ˈwɪnd fɑːm]
Windpark – Area of land with a cluster of → wind turbines.

wind turbine [ˈwɪnd ˈtɜːbaɪn]
Windrad – Tall structure with rotating blades that generates electricity.

windward side [ˈwɪndwə(r)d saɪd]
Luvseite – Side (mainly of a mountain) which faces the oncoming wind.

world trade [wɜːld treɪd]
Welthandel – International exchange of capital, goods, and services across borders.

Z

zenith [ˈzenɪθ]
Zenit – Assumed point that is vertically above the observer.

Phonetic symbols									
Vowels				*Dipthongs*		Consonants			
ɑː	arm	i	happy	aɪ	eye	ŋ	song	δ	these
ʌ	but	iː	easy	aʊ	our	r	red	θ	bathroom
e	desk	ɒ	orange	eə	air	s	sister	v	very
ə	an	ɔː	all	eɪ	take	z	zebra	w	what
ɜː	bird	ʊ	look	ɪə	here	ʒ	television	Stress	
æ	apple	u	January	ɔɪ	boy	dʒ	sausage	ˈ	Main stress in a word
ɪ	in	uː	boot	əʊ	go	ʃ	fresh	ˌ	Secondary stress in a word coming before the main stress.
				ʊə	your	tʃ	child		

M1 Physical map of Germany

M2 Physical map of Europe

M1 Physical map of the Earth

ASIA
AUSTRALIA
Siberia
Arctic Circle
Laptev Sea
East Siberian Sea
Kara Sea
Barents Sea
Novaya Zemlya
Kola Peninsula
Ural Mts.
Verkhoyansk Range
Lena
Stanovoy Range
Yablonovy Range
Sea of Okhotsk
Bering Sea
Aleutian Islands
Aleutian Trench
International Data Line
Yenisei
Irtysh
Novosibirsk
Moscow
Volga
Caspian Depression
Caspian Sea
Black Sea
Caucasus Mts.
5642
Mt. Elbrus
Altai
Turan
Tian Shan
Hindu Kush
Kunlun Shan
Gobi
Manchuria
Amur
Beijing
Sea of Japan
Korea
Tokyo
Japan
Japan Trench
10554
East China Sea
Huang He
Yangtse Kiang
Iranian Plateau
Tibetian Plateau
Mt. Everest
8846
Himalaya
Mesopotamia
Cairo
Arabia
Nile
Red Sea
Indus
Thar
Ganges
India
Mumbai
South China Plateau
Arabian Sea
Bay of Bengal
Bangkok
Mekong
South China Sea
Philippine Isles
10540
Marianas Trench
11034
Micronesia
Melanesia
Pacific Ocean
Ceylon
Malayan Archipelago
7450
New Guinea
Equator
Ethiopian Highlands
5895
Kilimanjaro
Dar es Salaam
East African Highlands
Madagascar
Drakensberg
Indian Ocean
Mid-Indian Ridge
Coral Sea
Great Sandy Desert
Great Victoria Desert
Great Artesian Basin
Great Dividing Range
Perth
Sydney
2230
Mt. Kosciusko
Tasman Sea
Kermadec Trench
Tasmania
New Zealand
Indian-Antarctic Ridge
Ridge
Antarctic Circle
Transantarctic Mountains
TICA
60° 80° 100° 120° 140° 160° 180° 160°
80° 66,5° 60° 40° 23,5° 20° 0° 20° 40° 60° 66,5° 80°

Country (2017)	Area in sq.km	population in mio.
Afghanistan	652864	34.66
Albania	28748	2.88
Algeria	2381741	40.60
Andorra	468	0.08
Angola	1246700	28.81
Antigua and Barbuda	442	0.1
Argentina	2780400	43.85
Armenia	29743	2.93
Australia	7692030	24.13
Austria	83882	8.75
Azerbaijan	86600	9.76
The Bahamas	13939	0.39
Bahrain	771	1.42
Bangladesh	147570	162.95
Barbados	430	0.29
Belarus	207600	9.51
Belgium	30528	11.35
Belize	22966	0.37
Benin	114763	10.87
Bhutan	38394	0.80
Bolivia	1098581	10.89
Bosnia and Herzegowina	51129	3.52
Botswana	582000	2.25
Brazil	8515767	207.65
Brunei	5765	0.42
Bulgaria	111002	7.13
Burkina Faso	274200	18.65
Burma /Myanmar	676577	52.89
Burundi	27834	10.52
Cambodia	181035	15.76
Cameroon	475650	23.44
Canada	9984670	36.29
Cape Verde	4033	0.54
Central African Republic	622984	4.56
Chad	1284000	14.45
Chile	756102	17.91
China	9596961	1386.62
Colombia	1141748	48.65
Comoros	1862	0.79
Congo	342000	5.13
Congo, D. R.	2344858	78.74
Costa Rica	51100	4.86
Ivory Coast	322463	23.70
Croatia	56594	4.17
Cuba	109884	11.47
Cyprus	9251	0.85
Czech Republic	78867	10.56
Denmark	42925	5.73
Djibouti	23200	0.94
Dominica	750	0.07
Dominican Republic	48661	10.65
East Timor	14604	1.02
Ecuador	257217	16.40
Egypt	1002000	95.69
El Salvador	21041	6.35
Equatorial Guinea	28051	1.22
Eritrea	117600	5.87
Estonia	45227	1.32
Ethiopia	1104300	102.40
Fiji	18272	0.90
Finland	338435	5.49
France	543965	66.87
Gabon	267668	1.98
The Gambia	11295	2.04
Georgia	69700	3.72
Germany	357409	82.67
Ghana	238537	28.20
Greece	131957	10.75
Grenada	345	0.11
Guatemala	108889	16.58
Guinea	245857	12.40
Guinea-Bissau	36125	1.82
Guyana	214969	0.77
Haiti	27750	10.85
Honduras	112492	9.11
Hungary	93024	9.82
Iceland	103000	0.34
India	3287263	1324.17
Indonesia	1913579	261.11
Iran	1648000	80.28
Iraq	435052	37.20
Ireland	69797	4.78
Israel	20991	8.55
Italy	302073	60.60
Jamaica	10991	2.88
Japan	377930	126.99
Jordan	89318	9.46
Kazakhstan	2724902	17.78
Kenya	591971	48.46
Kiribati	811	0.11
Korea, North	122762	25.37
Korea, South	99313	51.25
Kuwait	17818	4.05
Kyrgyzstan	199949	6.01
Laos	236800	6.76
Latvia	64573	1.96
Lebanon	10452	6.00
Lesotho	30355	2.20
Liberia	97754	4.61
Libya	1775500	6.29

Country (2017)	Area in sq. km	population in mio.
Liechtenstein	160	0.38
Lithuania	65 286	2.87
Luxembourg	2 586	0.58
Macedonia	25 713	2.01
Madagascar	587 041	24.89
Malawi	118 484	18.09
Malaysia	330 323	31.19
Maldives	300	0.42
Mali	1 240 192	17.99
Malta	316	0.44
Marshall Islands	181	0.05
Mauritania	1 030 700	4.30
Mauritius	2 040	1.25
Mexico	1 964 375	127.54
Micronesia	700	0.11
Moldova	33 846	3.55
Monaco	2	0.04
Mongolia	1 564 116	3.03
Montenegro	13 812	0.62
Morocco	446 550	35.28
Mozambique	799 380	28.83
Namibia	824 290	2.49
Nauru	21	0.01
Nepal	147 181	28.99
The Netherlands	41 526	17.02
New Zealand	270 534	4.18
Nicaragua	120 254	6.15
Niger	1 267 000	20.67
Nigeria	923 768	185.99
Norway	323 804	5.23
Oman	309 500	4.42
Pakistan	796 095	193.20
Palau	490	0.02
Panama	75 320	4.03
Papua New Guinea	462 840	8.08
Paraguay	406 752	6.72
Peru	1 285 216	31.77
Philippines	300 000	103.32
Poland	312 680	37.95
Portugal	92 226	10.32
Qatar	11 628	2.57
Romania	238 391	19.70
Russia	17 098 246	144.34
Rwanda	26 338	11.92
Saint Kits and Nevis	272	0.06
Saint Lucia	616	0.18
Saint Vincent and the Grenadines	398	0.12
Samoa	2 842	0.19
San Marino	61	0.03
São Tomé and Príncipe	1 001	0.20
Saudi Arabia	2 240 000	32.27
Senegal	196 722	15.41
Serbia	77 474	7.05
Seychelles	457	0.09
Sierra Leone	71 740	7.39
Singapore	719	5.61
Slovakia	49 035	5.43
Slovenia	20 273	2.06
Solomon Islands	28 896	0.60
Somalia	637 657	14.32
South Africa	1 219 602	55.91
Spain	505 944	46.44
Sri Lanka	65 610	21.20
Sudan	1 840 687	39.58
Surinam	163 820	0.56
Swaziland	17 363	1.34
Sweden	450 295	9.90
Switzerland	41 290	8.37
Syria	185 180	18.43
Tajikistan	142 600	8.73
Tanzania	947 303	55.57
Thailand	513 120	68.86
Togo	56 785	7.61
Tonga	748	0.10
Trinidad and Tobago	5 128	1.36
Tunisia	163 610	11.40
Turkey	783 562	79.52
Turkmenistan	488 100	5.66
Tuvalu	26	0.010
Uganda	241 550	41.49
Ukraine	603 500	45.00
United Arab Emirates	83 600	9.27
United Kingdom	248 532	65.64
United States of America	9 833 517	323.13
Uruguay	176 215	3.44
Uzbekistan	447 400	31.85
Vanuatu	12 189	0.27
Vatican City	0.4	0.008
Venezuela	912 050	31.57
Vietnam	330 967	92.70
Yemen	527 968	27.58
Zambia	752 612	16.59
Zimbabwe	390 757	16.15

New Countries since 2011

Country	Area in sq. km	population in mio.
Kosovo	10 887	1.82
South Sudan	619 745	12.23

M1 Countries of the world (2017)

Physical Map Symbols

Relief / landforms
- 5980 Mountain elevation (in metres)
- 3812 Other elevation (in metres)
- 6868 Sea depth
- Glacier
- Depression
- Tidal flats
- Coral reef

Inland waters
- River with waterfall
- Intermittent river
- Canal
- Lake, salt lake
- Moor, swamp / marsh
- Wadi, dry river bed
- Salt pan, salt lake

Transport
- Railway
- Major railway
- Furtherimportant rail link
- Tunnel
- Ferry
- Motorway
- Major road
- Major waterway

Settlements
Inhabitants
- Above 5 000 000
- 1 000 000 - 5 000 000
- 500 000 - 1 000 000
- 100 000 - 500 000
- 20 000 - 100 000
- Below 20 00
- Built up area

Administration
- National boundar
- Administrative boundary
- Disputed boundary
- Washington Capital
- Denver Principal place
- International airport

Individual map symbols
- Palace / stately home, c
- Church, monastry / convent / clo
- Grotto, cave
- Mountain pass
- Reservoir, dam wall
- Ruin, archeological site
- Oasis
- National park

Elevation (in metres)
- Above 5 000
- 3 000 - 5 000
- 1 500 - 3 000
- 1 000 - 1 500
- 500 - 1 000
- 200 - 500
- 100 - 200
- 0 - 100

Sea depth
- 0 - 200
- 200 - 2 000
- 2 000 - 4 000
- 4 000 - 6 000
- Below 6 000

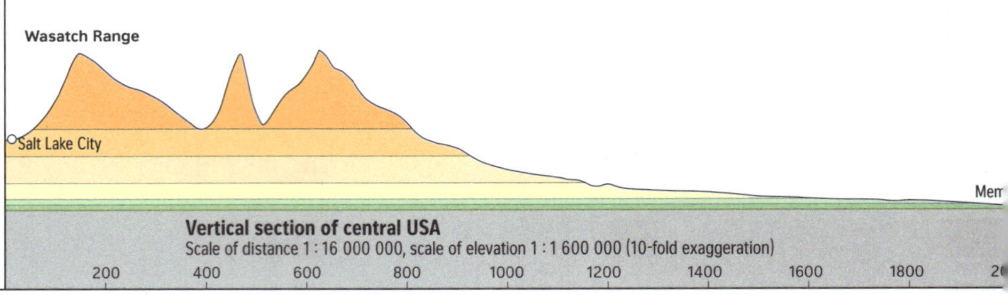

Vertical section of central USA
Scale of distance 1 : 16 000 000, scale of elevation 1 : 1 600 000 (10-fold exaggeration)

Taking Measurements on Maps

Maps represent natural features in reduced form. The scale names the amount of reduction between map distance and distance in nature (real distance). A scale of 1 : 100 000 means that 1 cm on the map equals 100 000 cm (1 km) on the earth's surface.

Example: Measure the distance Salt Lake City-Wichita on the map (by use of a set square) = 8 cm; read scale = 1 : 16 000 000.

Calculation: 8 cm on the map = 16 000 000 x 8 cm = 128 000 000 cm (1280 km) in reality.

Scale bar (comparison of frequently occuring scales):

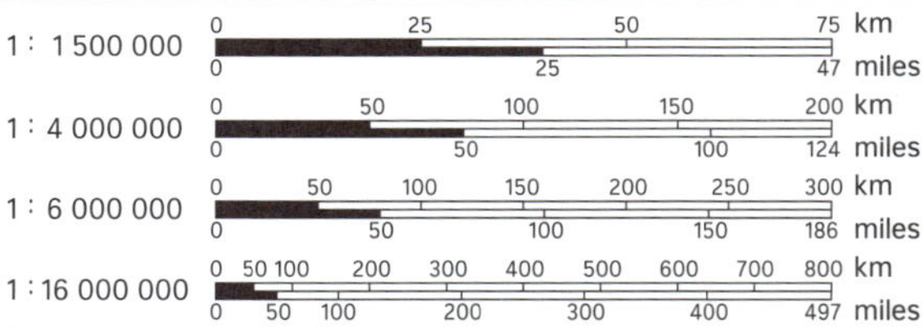

Note: Because of transferring the globe to a flat map, measurement mistakes may occur especially at smaller scales.

Orientation on Maps

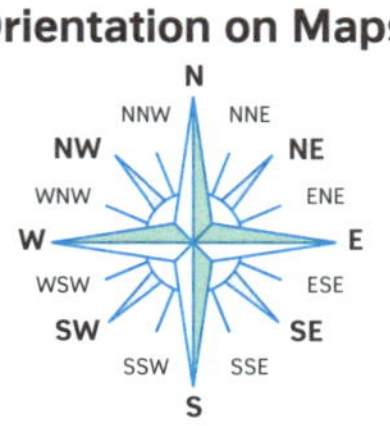

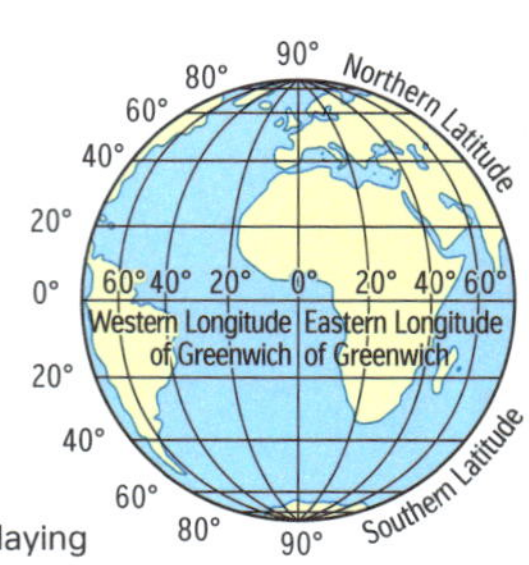

1. Indication of direction
Generally, all maps point towards the north. Exceptions are indicated by displaying a north arrow.
Indicate directions by using the wind rose.

2. Location of places
Every location on the surface of the globe can be fixed by its latitude and longitude figures.
Example:
Wichita 37°41′ North, 97°20′ West.

Atlas maps display geographical coordinates either on the map frame or along the edges of the map.

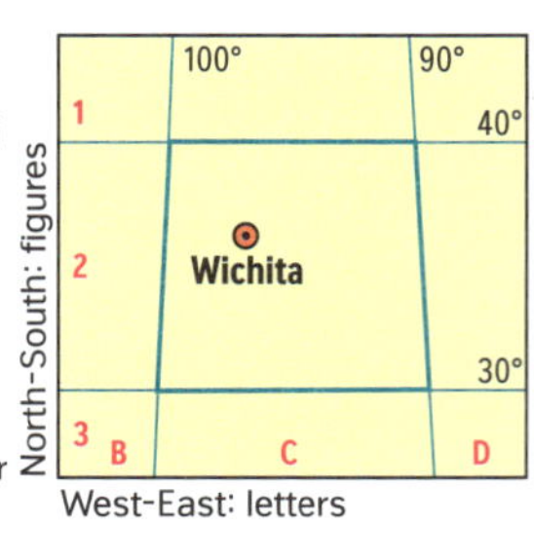

3. Searching for nam
Search for a name wi help of the alphabetic It can be found on th by using both page n and grid square. The square is made up of of latitude and longitu
Example: Wichita 10 Page 10, Grid squar

MAPS AND COUNTRIES

M1 Introduction to map reading 1: General key of physical maps

M2 Example, part of North America

Agriculture: The main crops and their relative values are shown by a combination of colours, patterns and symbols.

Spatial distribution of land cultivation

Farming on:

- Inferior soils
- Good soils
- Irrigated soils
- Pasture, meadows

Additional remarks specify the particular condition of agriculture in different climatic zones.

Map symbols

Cereal production / cultivation

- Wheat
- Maize (U.S.: corn)
- Rice
- Millet

On large-scale maps the map symbols appear more differentiated.

Sugar plants

- Sugar cane
- Sugar beet

Market gardening

- Bananas
- Date palms
- Hops
- Fruit and vegetables
- Citrus fruits
- Hazel-nuts — Special crops of great local significance, but low global distribution

Specialized crops

- Coffee
- Cacao
- Tobacco
- Tea
- Vine crops

Energy / commodity crops

Oil seeds

- Groundnuts
- Coconut palms
- Olives
- Palm oil
- Soy beans
- Sunflowers

Fibre crops

- Cotton
- Flax
- Jute
- Sisal

Other crops

- Rubber
- Cork trees
- Mulberry trees

The Crops are not quantified. Refer to the world map for shares contributed by individual states to world crop markets.

Mining: The index symbol is a rhombus ◇. The colours of the rhombus classify mineral resources, into main groups. Another geometric form within the rhombus signifies the specific mineral resource. According to their importance for the respective country as well as in global comparison, the symbols appear in three different sizes: ◇◇◇

Mineral resources									
Metal ores							Other minerals		Fossil resources (energy resources, raw material for the chemical industry)
Heavy metal ores						Light metals	Salts	Rocks and other minerals	
	Steel finishing metals	Other non-ferrous metals	High-grade metals / Precious metals	Energy resources					
Iron	Chrome	Antimony	Gold	Uranium		Aluminium raw material (e.g. bauxite)	Potassium salt	Diamonds	Hard coal
	Cobalt	Lead, zinc	Silver			Magnesium	Phosphates	Graphite	Brown coal
	Manganese	Copper	Platinum			Titanium	Potash	Limestone	(Mineral) Oil
	Molybdenum	Tin					Rock-salt	Gravel, sand	Natural gas
	Nickel							Kaolin clay	
	Vanadium	Mercury						Asbestos	
	Tungsten							Sulphur	

Industry: The index symbol is generally a circle ○, however, some industries have been assigned special symbols. The colours signify the branch of industry. The structure within the symbol mostly consists of easily depictable signs, which identify the branch of production within the industry.

According to their importance for the respective country as well as in global comparsion, the symbols appear in two different sizes: ○○. On all maps of larger scales the symbols have been differentiated further, in which case they appear indented.

Production of building material, glass, ceramics / pottery and porcelain / china

- Cement
- Glass
- Ceramics / pottery
- Porcelain / china

Metal production

- Iron and steel production
  - Steel production (oxygen steel converter and electric furnaces)
- Non-ferrous metal smelter
- Alumium smelter

Metal industry

- Rolling mill, foundry
  - Steel-girder construction
- Iron, tin, brass and metal products / goods

Machine, vehicle and equipment production

- Machine production
- Ship construction
- Motor vehicle construction
- Track vehicle construction
- Aerospace technology
- Missiles and space vehicles

Electrical engineering, information technology

- Electronics

Precision engineering, watches and clocks, optical products / goods

- Precision engineering
- Watch and clock making
- Optical products (laser technology)

Chemical industry

- Chemistry, synthetic material
- Biotechnology, pharmacy
- Rubber
- Petroleum refining
- Nuclear industry

Leather, textiles, clothes industry

- Leather
  - Leather goods
  - Shoes
- Textiles, clothing
  - Textiles
  - Clothing

Toy and jewellery industry, musical instruments

- Toys
- Musical instruments
- Jewellery

Wood / timber / lumber, paper and printing industry

- Wood / timber / lumber, furniture
  - Wood / timber / lumber processing
  - Furniture
- Cellulose, paper
  - Cellulose
  - Paper
- Printing works

Foods and semi-luxury foods industry

- Fish processing
- Meat processing
- Beverage production
- Milk processing
- Tobacco production
- Sugar refinery

Energy: The size of the symbols for power stations varies with the capacity to produce electricity. The smallest symbols generally represent an anual production of less than 1000 MW.

- Hydroelectric power station (damned or undamned water, pumpe water)
- Thermal power station (coal, oil, natural gas)
- Nuclear power station
- Solar power station (photovoltaics)
- Geothermal power station

Services: The index symbol for services is a hexagon ⬡. Service centres are shown and categorized according to their importance.

Service centre (scale of 1 : 16 000 000 and smaller)

- Of international importance (world cities with a comprehensive service sector)
- Of super-regional importance (large cities with a large service sector)
- Of regional importance (cities with selected services)
- Tourism location or region of international importance

Large-scale maps differentiate among services: only typical service sub-sectors are shown for the place in question.

Services (scale 1: 6 000 000 and larger)

Administration, supplies (in general evenly spaced, therefore only shown at detailed scales)

- Government offices, other administration, legal offices
- Supply and disposal
- Health services, retirement homes

Education, media

- University or other
- Media (television, press, radio)
- Culture (theatre, opera, museum)

Finances, trade

- Financial centre
- Commercial enterprise
- Service, consulting
- Trade fairs, exhibitions

Transport, logistics (turnover of goods, travel)

- Airport
- Port
- Logistics (storage / warehouse, haulage company)
- Military base

On economic maps the central places are graded according to their number of inhabitants.

- More than 1 000 000 inhabitants
- 100 000 to 1 000 000 inhabitants
- Below 100 000 inhabitants

Infrastructure:

The land-based transport of people, goods and merchandise between centres uses the following traffic routes.

Traffic and transport

- Railroads
- Motorway (U.S.: interstate
- Long-distance roads
- Oil pipeline (wi oil terminal)
- Gas pipeline

13902

M1 Introduction to map reading 2: General key of thematic maps

M2 Example, part of North America

Sources

A1PIX - Your Photo Today, Ottobrunn: 4 (Hans Strand); Cagle Cartoons, Santa Barbara, CA: 30; Druwe & Polastri, Cremlingen/Weddel: 7 ff Skills Hack; fotolia.com, New York: 75 (ADESIGN); Getty Images, München: Titel li. (John Wolfe-Barry), Titel m. (Shaun Egan/JAI), 4 (AFP PHOTO/Jewel Samad); Google Earth: 7 (Image ©2017 TerraMetrics), 15; iStockphoto.com, Calgary: 40 (Foto: NNehring); Kirch, Peter, Koblenz: 44; Mittal Steel Company, Chicago: 14; Müller, Rolf, Ølstykke: 12; NASA/GSFC, Houston/Texas: 5; Picture-alliance Frankfurt/M.: 4 (EPA/Adi Weda); Rieger, Anja, Backnang: 28 Stifte; Schaller, G., Braunschweig: 50; Schönauer-Kornek, Sabine, Wolfenbüttel: 28; Shutterstock.com, New York: Titel re.; wikimedia.commons: 74.